Oxford **Mathematics**
Primary Years Programme
3

W0036620

Contents

OXFORD
UNIVERSITY PRESS
AUSTRALIA & NEW ZEALAND

5367 is the same as:

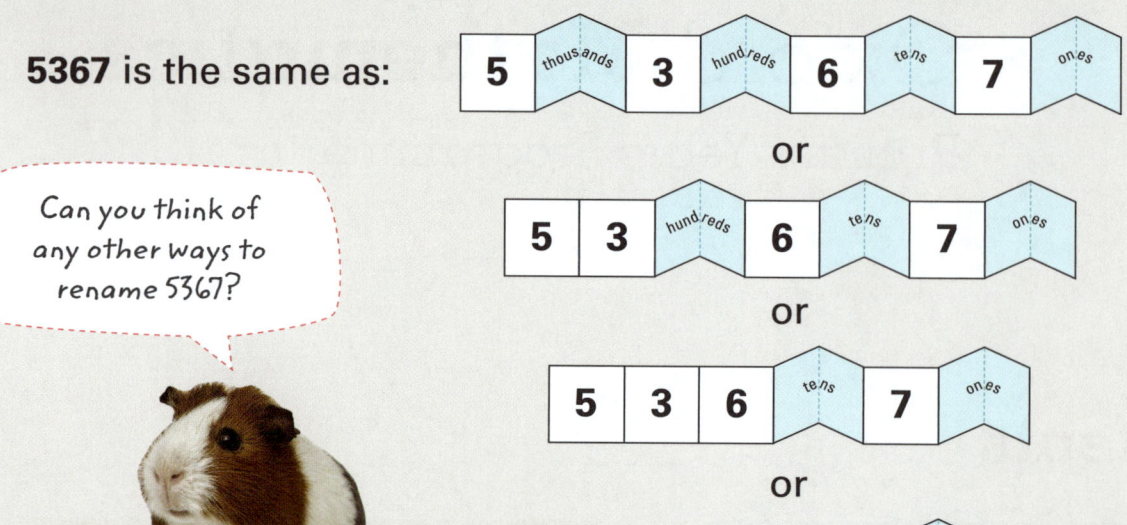

Can you think of any other ways to rename 5367?

Guided practice

1 Show these numbers on the number expanders.

a 2431

b 8276

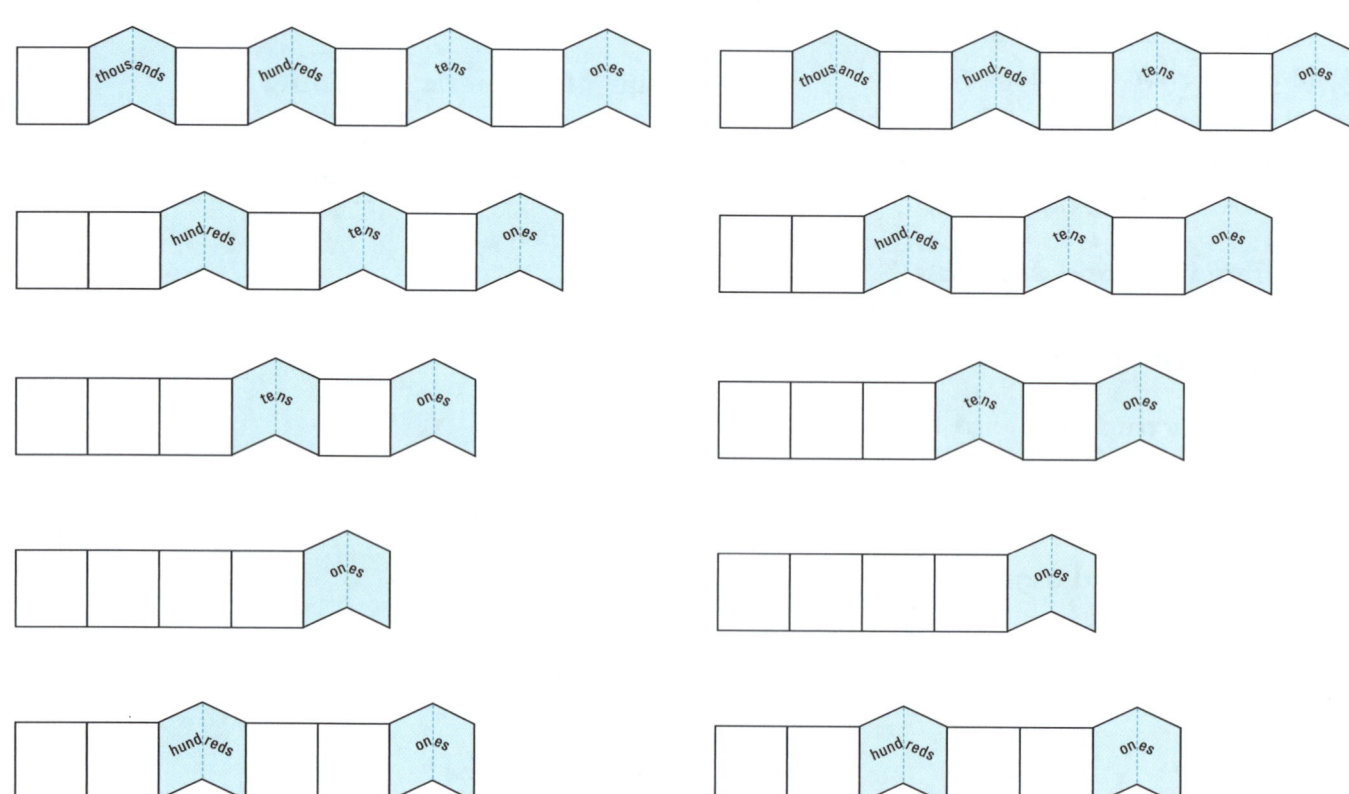

OXFORD UNIVERSITY PRESS

Independent practice

Write each number:

1 in words.

a 4568 _____

_____

_____

b 8043 _____

_____

_____

c 7109 _____

_____

_____

2 on the place value chart.

Th	H	T	O

How do the numbers in words connect with the place value chart?

3 How many?

a

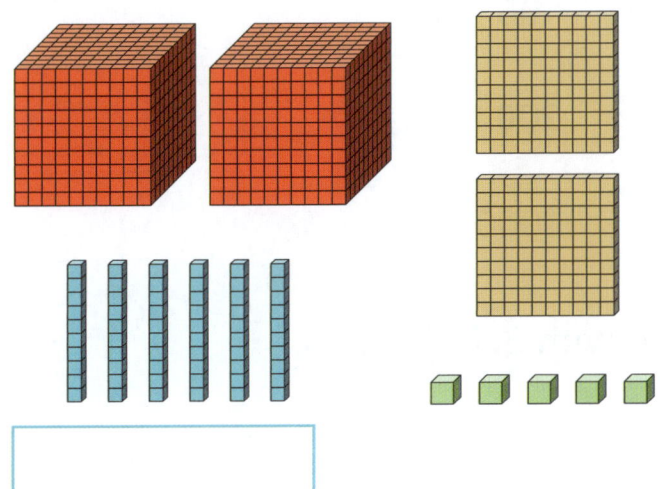

b

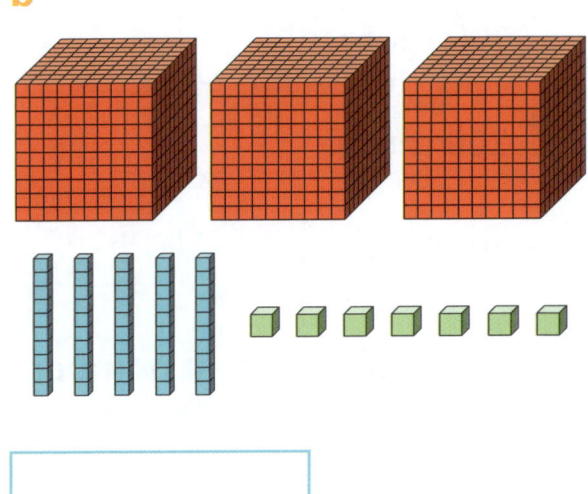

4 Rewrite the number of people in the table from largest to smallest.

WORLD PARTICIPATION RECORDS

Event number	Event	Number of people		Event number	Number of people
1	Most people dressed as Smurfs	4891			
2	Largest Riverdance line	1693			
3	Largest Thai dance	5255			
4	Largest umbrella dance	1688			
5	Largest lion dance	3971			
6	Largest scarecrow display	3812			

5 Make the largest number possible with 1, 7, 8 and 0.

6 Use the number from question 5 to find:

a 10 more.

b 10 less.

c 20 more.

d 20 less.

e 100 more.

f 100 less.

g 200 more.

h 200 less.

i 1000 more.

j 1000 less.

7 Make the smallest number possible with 3, 8, 2 and 3.

OXFORD UNIVERSITY PRESS

Extended practice

1 Write on the expander, then complete the sum.

a 3790 =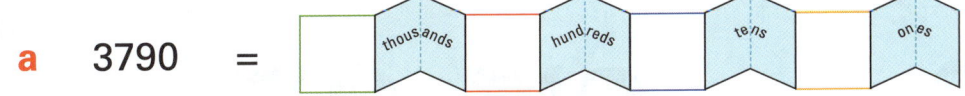

3790 = ☐ + ☐ + ☐ + ☐

b 8052 =

8052 = ☐ + ☐ + ☐

c 24 160 =

24 160 = ☐ + ☐ + ☐

2 Circle the number in which:

a 4 has the greatest value.   3472   6324   4012

b 9 has the smallest value.   6889   3914   1900

c 1 has the greatest value.   5217   1024   9199

d 5 has the smallest value.   19 875   2536   6851

3 **a** Write the largest and the smallest 4-digit number possible with 7 in the tens column.

b Write the largest and the smallest 4-digit number possible with 4 in the hundreds column.

Even numbers can be grouped into 2s.

Odd numbers cannot.

8

7

What is an odd number? What other meaning does the word **odd** have?

Guided practice

1 Circle groups of 2, and then colour if the total is odd or even.

a

| Odd | Even |

b

| Odd | Even |

c

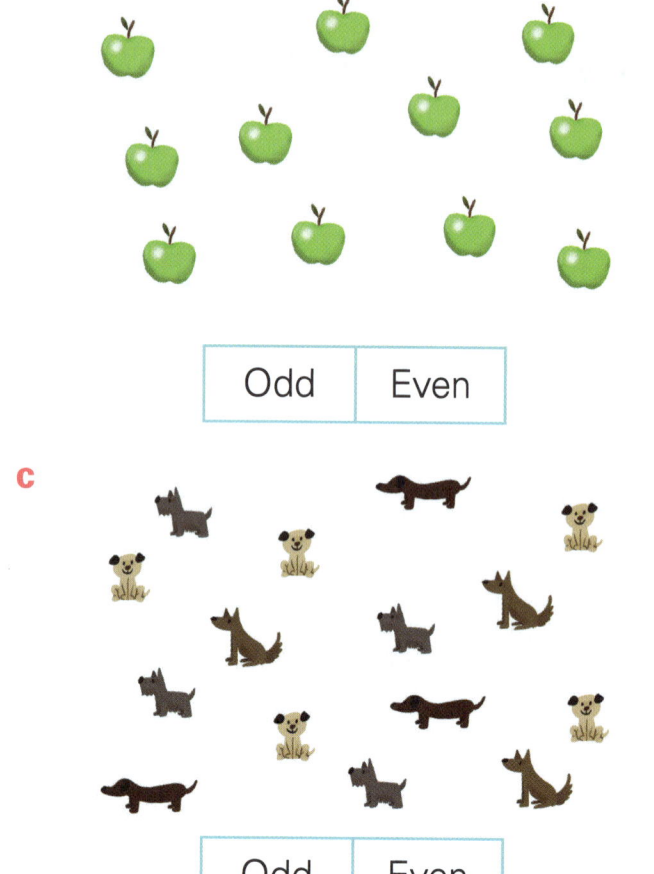

| Odd | Even |

d

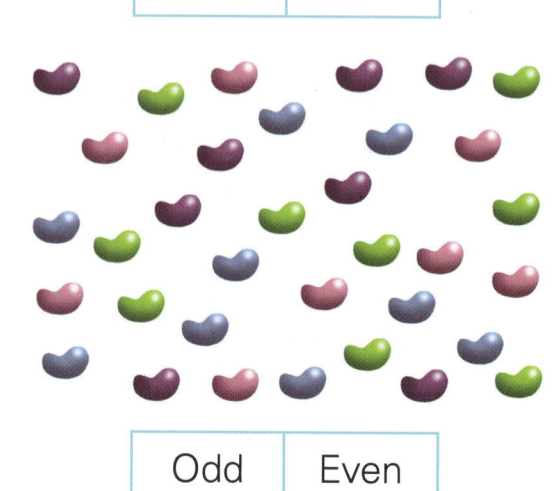

| Odd | Even |

OXFORD UNIVERSITY PRESS

1 Draw on the ten frames, and then choose if the numbers are odd or even.

a   17

Odd
Even

b   26

Odd
Even

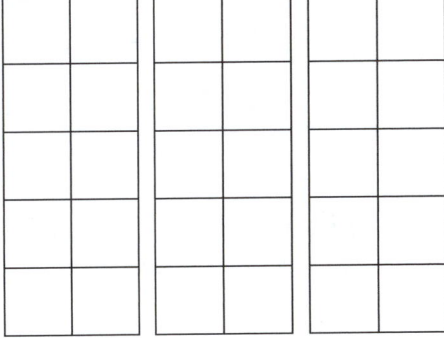

c   28

Odd
Even

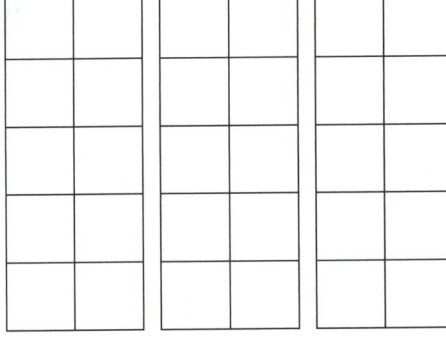

d   14

Odd
Even

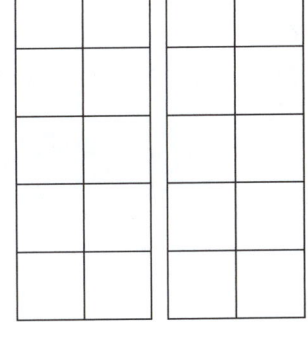

e   25

Odd
Even

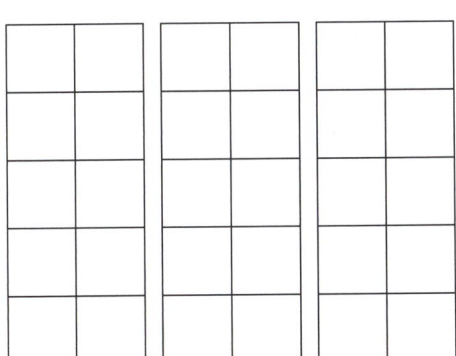

f   15

Odd
Even

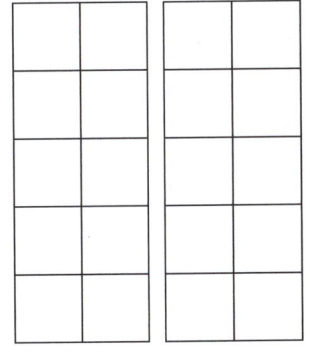

2 Finish the number patterns.

a   Odd:

21		25	27			33		

b   Even:

44	46			52			58	

c   Even:

20	24	28			40			52

3

31	32	33	34	35	36	37	38	39	40
41	42	43	44	45	46	47	48	49	50
51	52	53	54	55	56	57	58	59	60

a Circle all the even numbers in **red**.

b Circle all the odd numbers in **blue**.

c What digits can even numbers end in?

☐ ☐ ☐ ☐ ☐

d What digits can odd numbers end in?

☐ ☐ ☐ ☐ ☐

Which place value column tells you if a number is odd or even?

4 Rewrite the numbers in the correct column.

Odd	Even

76   143   258

103   575   1974

1361   3870   5002

867   9998   9999

5 Odd or even?

a The number of fingers on one hand _____

b On two hands _____

c The number of wheels on one car _____

d On two cars _____

OXFORD UNIVERSITY PRESS

1 Add the pairs of even numbers.

a   6 + 2 = ☐   b   14 + 10 = ☐   c   28 + 8 = ☐

d   All the answers are: | Odd | Even |

2 Add the pairs of odd numbers.

a   5 + 3 = ☐   b   11 + 17 = ☐   c   21 + 9 = ☐

d   All the answers are: | Odd | Even |

3 Add the pairs of even and odd numbers.

a   4 + 5 = ☐   b   12 + 15 = ☐   c   20 + 19 = ☐

d   All the answers are: | Odd | Even |

4 Add the pairs of odd and even numbers.

a   5 + 6 = ☐   b   17 + 10 = ☐   c   23 + 14 = ☐

d   All the answers are: | Odd | Even |

5 Will the answer be odd or even?

a   24 + 56 | Odd | Even | b   45 + 38 | Odd | Even |

c   72 + 93 | Odd | Even | d   88 + 66 | Odd | Even |

e   97 + 75 | Odd | Even | f   51 + 94 | Odd | Even |

One-digit numbers can help you add bigger numbers.

If you know:	You also know:	Or:
6 + 3 = 9	16 + 3 = 19	6 + 13 = 19

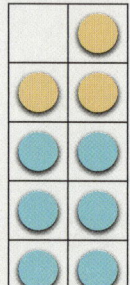 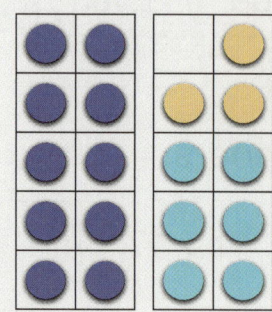 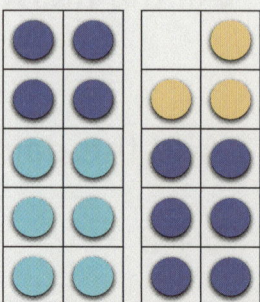

What would 16 + 13 be?

Guided practice

1 Find the answers.

a 4 + 3 = ☐ and 14 + 3 = ☐

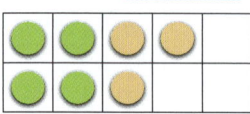 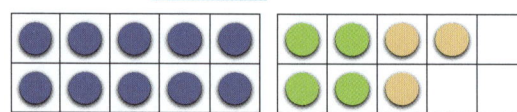

b 2 + 6 = ☐ and 12 + 6 = ☐

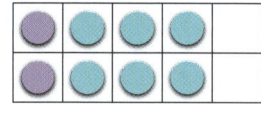 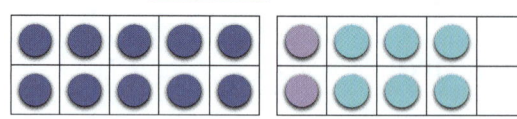

c 8 + 2 = ☐ and 8 + 12 = ☐

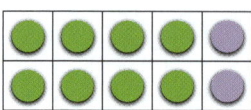 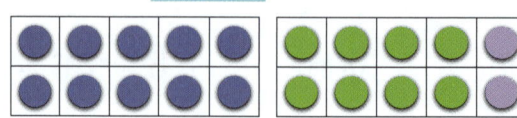

d 1 + 4 = ☐ and 21 + 4 = ☐

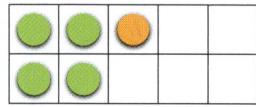 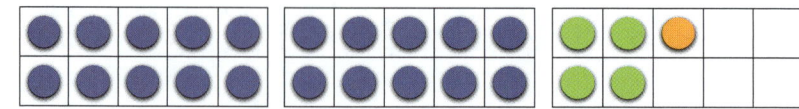

OXFORD UNIVERSITY PRESS

Independent practice

1 Extend the number facts to solve.

a 2 + 7 = [ ]   and   22 + 7 = [ ]

b 5 + 3 = [ ]   and   5 + 13 = [ ]

c 2 + 4 = [ ]   and   12 + 14 = [ ]

d 1 + 8 = [ ]   and   31 + 8 = [ ]

e 6 + 4 = [ ]   and   6 + 34 = [ ]

What other mental addition strategies could you use?

2 Use doubles facts to solve.

a If 3 + 3 = 6, then 30 + 30 = [ ] .

b If 4 + 4 = [ ] , then 40 + 40 = [ ] .

c If 5 + 5 = [ ] , then 50 + 50 = [ ] .

d If 2 + 2 = [ ] , then [ ] + [ ] = 40.

e If 8 + 8 = [ ] , then [ ] + [ ] = 160

f If 1 + 1 = [ ] , then 100 + 100 = [ ] .

g If 6 + 6 = [ ] , then 600 + 600 = [ ] .

h If 7 + 7 = [ ] , then 700 + 700 = [ ] .

3 Split into 10s and 1s to add.

a 23 + 12 = [30] + [5] = [ ]

b 26 + 31 = [ ] + [ ] = [ ]

c 45 + 42 = [ ] + [ ] = [ ]

d 34 + 55 = [ ] + [ ] = [ ]

e 43 + 27 = [ ] + [ ] = [ ]

When adding in your head, it's easier if you can make pairs that equal a 10.

4 Rearrange the numbers to make them easier to add.

a 6 + 7 + 4 = [6] + [4] + [7] = [ ]

b 5 + 4 + 25 = [ ] + [ ] + [ ] = [ ]

c 17 + 2 + 4 + 3 = [ ] + [ ] + [ ] + [ ] = [ ]

d 3 + 11 + 2 + 19 = [ ] + [ ] + [ ] + [ ] = [ ]

5 Solve using a mental addition strategy of your choice.

a 90 + 90 = [ ]

b 46 + 52 = [ ]

c 4 + 37 = [ ]

d 17 + 8 + 3 + 12 = [ ]

e 21 + 68 = [ ]

f 500 + 500 = [ ]

g 61 + 17 = [ ]

h 14 + 30 + 6 = [ ]

OXFORD UNIVERSITY PRESS

The table below shows how many people went on each ride at an amusement park in a one-hour period.

Ride	Roller coaster	Carousel	Big slide	Haunted house	Ferris wheel	Tea cups	Giant drop	Dodgem cars
Number of people	23	8	7	54	135	12	39	221

1 Write the numbers in the easiest adding order to find how many people went on:

a the carousel, big slide and tea cups.

☐ + ☐ + ☐ = ☐

b the big slide, tea cups and roller coaster.

☐ + ☐ + ☐ = ☐

c the carousel, dodgem cars and giant drop.

☐ + ☐ + ☐ = ☐

---

2 Add in your head to find how many people went on:

a the haunted house and the giant drop.

☐ + ☐ = ☐

b the dodgem cars and the roller coaster.

☐ + ☐ = ☐

c the Ferris wheel and the haunted house.

☐ + ☐ = ☐

d the dodgem cars and the Ferris wheel.

☐ + ☐ = ☐

e the roller coaster, the carousel, the tea cups and the big slide.

☐ + ☐ + ☐ + ☐ = ☐

Jump strategy for addition

Start with the larger number. Add the 10s, and then the 1s.

22 + 23

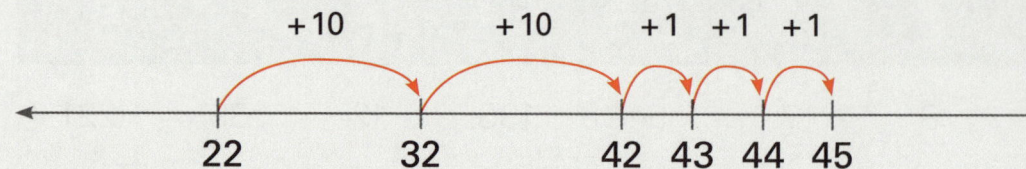

+10   +10   +1 +1 +1

22   32   42 43 44 45

Where would you start if you were adding 2 hundreds numbers?

Guided practice

1   Use the jump strategy to solve.

a   16 + 21 = ⬚

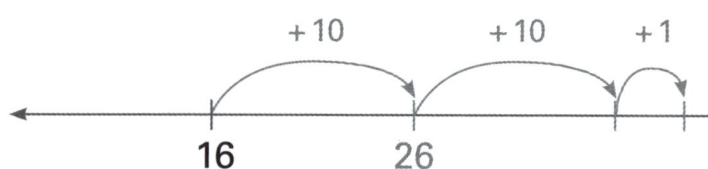

+ 10   + 10   + 1

16   26

b   35 + 24 = ⬚

35

c   146 + 33 = ⬚

146

OXFORD UNIVERSITY PRESS

Independent practice

1 Use the jump strategy.

a 72 + 25 = ☐

b 112 + 57 = ☐

c 231 + 63 = ☐

d 320 + 41 = ☐

e 25 + 414 = ☐

Vertical addition

125 + 273

Add the ones

H	T	O
1	2	5
+ 2	7	3
		8

Then the tens

H	T	O
1	2	5
+ 2	7	3
	9	8

Then the hundreds

H	T	O
1	2	5
+ 2	7	3
3	9	8

Guided practice

1 Start with the ones to solve.

a

H	T	O
	4	4
+	5	2

b

H	T	O
1	0	1
+	6	7

c

H	T	O
2	5	3
+ 1	3	4

d

H	T	O
4	1	0
+ 3	3	6

e

H	T	O
6	3	7
+ 2	4	2

f

H	T	O
8	1	4
+ 1	8	2

g

H	T	O
	5	3
+ 4	2	1

h

H	T	O
5	5	5
+ 3	3	3

i

H	T	O
8	0	2
+ 1	0	7

OXFORD UNIVERSITY PRESS

Remember to line the numbers up in their place value columns.

1 Rewrite as vertical addition and solve.

a 28 + 31   **b** 63 + 35   **c** 46 + 22

+   +   +

d 358 + 421   **e** 480 + 217   **f** 891 + 206

+   +   +

---

2 Write as vertical addition and solve.

a Serena counted 328 cars on the way to school and 451 cars on the way home.

How many did she count altogether?

b Arjun drove 236 km on Saturday and 603 km on Sunday.

How far did he travel on the weekend?

Extended practice

> How are the jump strategy and vertical addition similar?

1 Use the jump strategy.

a   375 + 427 = [   ]

⟵————————————————————————⟶

b   681 + 242 = [   ]

⟵————————————————————————⟶

- -

2 Use vertical addition.

a

```
   1375
+   413
_____

_____
```

b

```
   2517
+  1002
_____

_____
```

c

```
   6350
+  1237
_____

_____
```

- -

3 Choose a strategy to find the answer.

324 + 543 = [   ]

[   ]

OXFORD UNIVERSITY PRESS

One-digit numbers can help you to subtract bigger numbers.

If you know:	You also know:	Or:
$7 - 2 = 5$	$17 - 2 = 15$	$27 - 2 = 25$

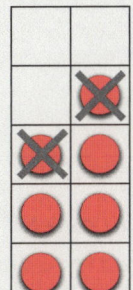 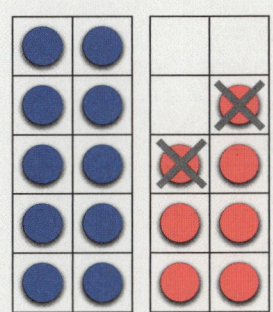 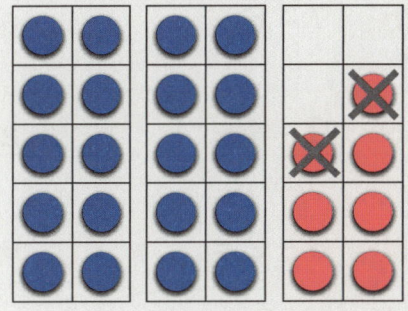

Guided practice

What other subtraction strategies could you use?

1 Find the answers.

a $9 - 6 = \boxed{}$ and $19 - 6 = \boxed{}$

 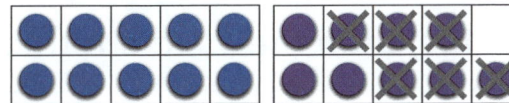

b $8 - 1 = \boxed{}$ and $18 - 1 = \boxed{}$

 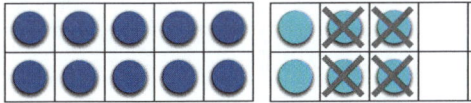

c $6 - 4 = \boxed{}$ and $16 - 4 = \boxed{}$

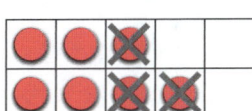 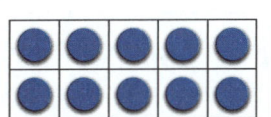

d $7 - 3 = \boxed{}$ and $27 - 3 = \boxed{}$

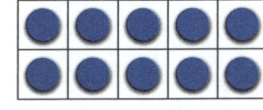 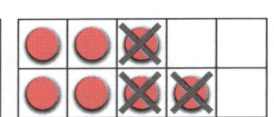

1 Extend the number facts to solve.

a   5 − 3 = [ ]   and   15 − 3 = [ ]

b   7 − 6 = [ ]   and   27 − 6 = [ ]

c   9 − 4 = [ ]   and   19 − 4 = [ ]

d   8 − 2 = [ ]   and   28 − 2 = [ ]

e   6 − 3 = [ ]   and   36 − 3 = [ ]

f   4 − 2 = [ ]   and   84 − 2 = [ ]

g   7 − 4 = [ ]   and   97 − 4 = [ ]

Can you extend the number facts to work out 115 − 3 in your head?

2 Take away the 10s, then the 1s to subtract.

a   35 − 13 = [35] − [10] − [3] = [ ]

b   48 − 15 = [ ] − [ ] − [ ] = [ ]

c   52 − 21 = [ ] − [ ] − [ ] = [ ]

d   67 − 34 = [ ] − [ ] − [ ] = [ ]

e   96 − 25 = [ ] − [ ] − [ ] = [ ]

f   124 − 13 = [ ] − [ ] − [ ] = [ ]

g   389 − 57 = [ ] − [ ] − [ ] = [ ]

OXFORD UNIVERSITY PRESS

Subtracting to ten is a good strategy because it is easier to take away from a ten.

3 Subtract to a ten to solve.

a 26 – 8 = [26] – [6] – [2] = [ ]

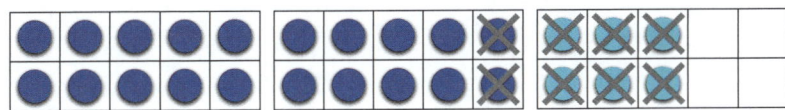

b 32 – 7 = 32 – [ ] – [ ] = [ ]

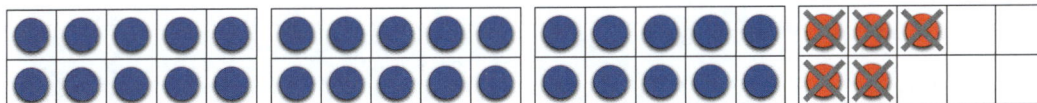

c 35 – 9 = 35 – [ ] – [ ] = [ ]

d 21 – 6 = 21 – [ ] – [ ] = [ ]

e 43 – 5 = 43 – [ ] – [ ] = [ ]

f 64 – 7 = [ ] – [ ] – [ ] = [ ]

g 76 – 9 = [ ] – [ ] – [ ] = [ ]

h 145 – 8 = [ ] – [ ] – [ ] = [ ]

Extended practice

1 Use extended number facts to solve.

a 7 – 5 = ☐   and   70 – 50 = ☐

b 9 – 2 = ☐   and   90 – 20 = ☐

c 8 – 4 = ☐   and   80 – 40 = ☐

d 4 – 2 = ☐   and   400 – 200 = ☐

e 6 – 5 = ☐   and   600 – 500 = ☐

2 Solve in your head.

a Baxter had 28 balloons. 14 of them popped. How many are left?

b 94 children were at the bus stop. 35 got on the first bus. How many are left?

c Eloise made 164 cups of lemonade. She sold 23 cups in the first hour. How many cups does she still need to sell?

d Brittany picked up 132 pieces of rubbish at clean up day. Ashley arrived late and only picked up 8. How many more than Ashley did Brittany pick up?

OXFORD UNIVERSITY PRESS

Jump strategy for subtraction

Take away the 10s, and then the 1s.

48 − 24

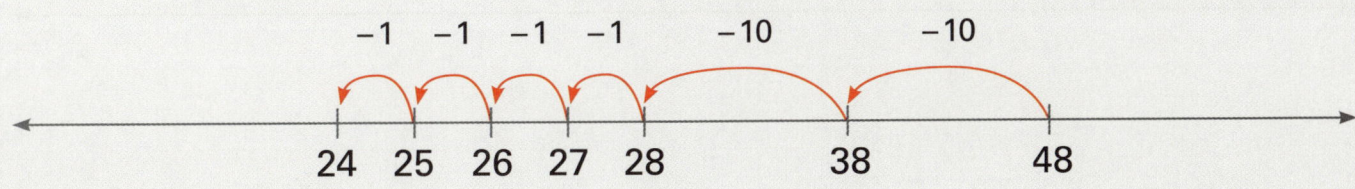

What would you take away first if you were using the jump strategy for hundreds numbers?

Guided practice

1 46 − 23 = ☐

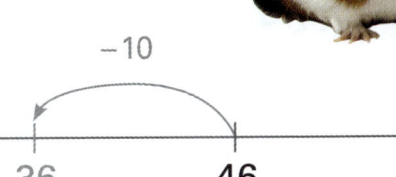

−10

36   46

---

2 58 − 35 = ☐

58

---

3 263 − 41 = ☐

263

Independent practice

1 Use the jump strategy.

a 98 − 34 = ☐

b 360 − 43 = ☐

c 798 − 51 = ☐

d 598 − 125 = ☐

e 372 − 203 = ☐

OXFORD UNIVERSITY PRESS

Vertical subtraction

564 − 342

Subtract the ones

H	T	O
5	6	4
− 3	4	2
		2

Then the tens

H	T	O
5	6	4
− 3	4	2
	2	2

Then the hundreds

H	T	O
5	6	4
− 3	4	2
2	2	2

Guided practice

1 Start with the ones to solve.

a

T	O
3	7
− 1	4

b

H	T	O
4	6	8
−	2	1

c

H	T	O
8	7	7
− 3	0	2

d

H	T	O
9	4	3
− 2	1	1

e

H	T	O
6	4	9
− 4	2	6

f

H	T	O
7	1	8
− 2	1	4

g

H	T	O
5	0	1
− 3	0	1

h

H	T	O
9	6	0
− 2	3	0

i

H	T	O
8	8	8
− 5	5	5

Remember to line the numbers up in their place value columns.

1 Rewrite as vertical subtraction and solve.

a 27 – 14

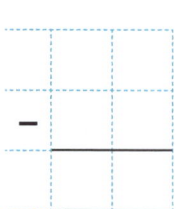

b 53 – 31

c 86 – 36

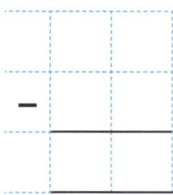

d 173 – 162

e 797 – 493

f 891 – 206

2 Write as vertical subtraction and solve.

a Betty the baker made 98 cupcakes. She sold 57 of them.

How many are left?

b Suresh had 645 new emails. He opened 414 of them.

How many are still unread?

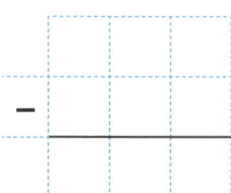

OXFORD UNIVERSITY PRESS

1 Solve using the jump strategy.

a 742 − 216 = [ ]

←――――――――――――――――――――――――→

b 628 − 343 = [ ]

←――――――――――――――――――――――――→

2 Solve using vertical subtraction.

a

Th	H	T	O
	5	7	2	6
−		5	1	2

b

Th	H	T	O
3	8	6	7
− 1	2	0	5

c

Th	H	T	O
7	5	3	1
− 5	0	2	0

3 Choose a strategy to find the answer.

967 − 452 = [ ]

Subtraction undoes addition.

$$10 + 5 = 15$$

$$15 - 5 = 10$$

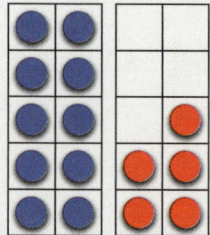 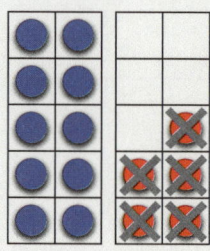

Inverse means opposite.

Guided practice

1 Use the addition facts to complete the subtraction facts.

a $7 + 5 = 12$ $12 - 5 = \boxed{}$

b $24 + 9 = 33$ $33 - 9 = \boxed{}$

c $38 + 7 = 45$ $45 - 7 = \boxed{}$

2 Use the subtraction facts to complete the addition facts.

a $9 - 3 = 6$ $3 + 6 = \boxed{}$

b $27 - 8 = 19$ $19 + 8 = \boxed{}$

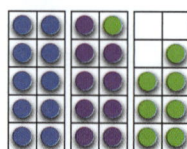

c $43 - 7 = 36$ $36 + 7 = \boxed{}$

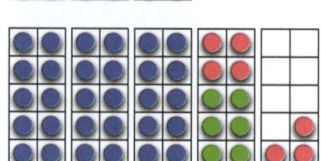

OXFORD UNIVERSITY PRESS

Fact families are sets of number facts that are related.

1 Complete the fact families.

a

10

6   4

6 + 4 = 10   4 + 6 = ☐

10 − 6 = ☐   10 − 4 = ☐

b

24

17   7

17 + 7 = 24   7 + ☐ = 24

24 − ☐ = 17   24 − ☐ = 7

c

29

17   12

☐ + 12 = 29   12 + ☐ = 29

29 − 17 = 12   29 − ☐ = 17

d

48

40   8

☐ + ☐ = 48   ☐ + ☐ = 48

48 − ☐ = 40   48 − ☐ = 8

e

82

45   37

45 + ☐ = 82   37 + ☐ = 82

82 − ☐ = 45   82 − ☐ = 37

f

126

26   100

100 + ☐ = 126   26 + ☐ = 126

126 − ☐ = 100   126 − ☐ = 26

2 Use each set of numbers to make 2 addition and 2 subtraction equations.

a

14	17	31

☐ + ☐ = ☐

☐ + ☐ = ☐

☐ − ☐ = ☐

☐ − ☐ = ☐

b

32	46	78

☐ + ☐ = ☐

☐ + ☐ = ☐

☐ − ☐ = ☐

☐ − ☐ = ☐

c

15	48	33

☐ + ☐ = ☐

☐ + ☐ = ☐

☐ − ☐ = ☐

☐ − ☐ = ☐

d

55	39	16

☐ + ☐ = ☐

☐ + ☐ = ☐

☐ − ☐ = ☐

☐ − ☐ = ☐

e

97	70	167

☐ + ☐ = ☐

☐ + ☐ = ☐

☐ − ☐ = ☐

☐ − ☐ = ☐

f

278	143	135

☐ + ☐ = ☐

☐ + ☐ = ☐

☐ − ☐ = ☐

☐ − ☐ = ☐

You can use addition to check your subtraction answers and subtraction to check your addition answers.

OXFORD UNIVERSITY PRESS

Extended practice

The compensation strategy uses rounding and inverse operations to make numbers easier to work with.

For example: 45 + 39 is the same as 45 + 40 − 1 = 84.

Add 1 to round 39 to 40, and then subtract 1 to undo the addition.

1 Solve these additions using rounding and subtraction.

a 34 + 28 is the same as 34 + ☐ − 2 = ☐ .

b 26 + 29 is the same as 26 + ☐ − 1 = ☐ .

c 53 + 49 is the same as 53 + ☐ − ☐ = ☐ .

d 45 + 27 is the same as 45 + ☐ − 3 = ☐ .

e 54 + 17 is the same as 54 + ☐ − ☐ = ☐ .

2 Multiplication and division are also inverse operations. Finish the fact families.

a

$2 \times 10 = 20$ $10 \times 2 = $ ☐

$20 \div 2 = $ ☐ $20 \div 10 = $ ☐

b

$4 \times 12 = 48$ $12 \times $ ☐ $ = $ ☐

$48 \div 4 = $ ☐ $48 \div $ ☐ $ = $ ☐

c

$8 \times 7 = 56$ $7 \times $ ☐ $ = $ ☐

$56 \div 7 = $ ☐ $56 \div 8 = $ ☐

d

$9 \times $ ☐ $ = 99$ $11 \times $ ☐ $ = 99$

$99 \div $ ☐ $ = $ ☐ $99 \div $ ☐ $ = $ ☐

3 Use inverse operations to solve.

a

$73 \times 5 = 365$ $365 \div $ ☐ $ = 5$

b

$1532 − 845 = 687$ $687 + 845 = $ ☐

Multiplication and division are inverse operations.

1 group of 2 is 2.

2 shared between 1 is 2.

2 groups of 2 are 4.

4 shared between 2 is 2.

What other inverse operations do you know?

Guided practice

1 Use the multiplication facts to complete the division facts.

a 3 groups of 5 = 15. 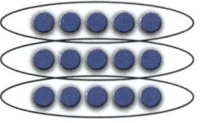 15 shared between 3 is ☐.

b 6 groups of 2 = 12.  ☐ shared between 6 is ☐.

c 4 groups of 7 = 28. 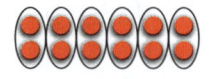 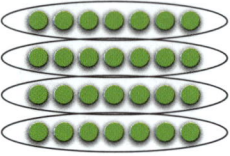 ☐ shared between 4 is ☐.

---

2 Use the division facts to complete the multiplication facts.

a 9 shared between 3 is 3.  ☐ groups of 3 = 9.

b 16 shared between 8 is 2.  ☐ groups of 2 = ☐.

c 18 shared between 3 is 6. 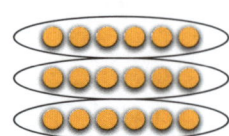 ☐ groups of 6 = ☐.

OXFORD UNIVERSITY PRESS

We use the × sign for "groups of" or multiplication, and the ÷ sign for sharing or division.

Independent practice

1 Make turnaround multiplication facts to match each array.

a

| 3 | × | 4 | = | 12 |

| 4 | × | ☐ | = | ☐ |

b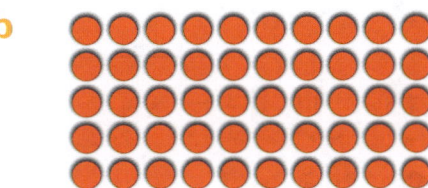

| ☐ | × | ☐ | = | ☐ |

| ☐ | × | ☐ | = | ☐ |

c

| ☐ | × | ☐ | = | ☐ |

| ☐ | × | ☐ | = | ☐ |

d

| ☐ | × | ☐ | = | ☐ |

| ☐ | × | ☐ | = | ☐ |

2 Complete the fact families.

a $3 \times 9 = 27$

☐ × ☐ = 27

27 ÷ ☐ = ☐

27 ÷ ☐ = ☐

b $10 \times 2 = 20$

☐ × ☐ = 20

20 ÷ ☐ = ☐

20 ÷ ☐ = ☐

c $8 \times 5 = $ ☐

5 × ☐ = ☐

☐ ÷ 5 = ☐

☐ ÷ ☐ = 5

d $7 \times 10 = $ ☐

☐ × ☐ = ☐

☐ ÷ ☐ = ☐

☐ ÷ ☐ = ☐

3 Complete the multiplication facts.

a ● ● ● 1 × 3 = ☐

b ● ● ● 2 × 3 = ☐

c ● ● ● 3 × 3 = ☐

d ● ● ● 4 × 3 = ☐

e ● ● ● 5 × 3 = ☐

f ● ● ● 6 × 3 = ☐

g ● ● ● 7 × 3 = ☐

h ● ● ● 8 × 3 = ☐

i ● ● ● 9 × 3 = ☐

j ● ● ● 10 × 3 = ☐

4 Now, write a matching division fact.

a 3 ÷ ☐ = ☐

b 6 ÷ ☐ = ☐

c ☐ ÷ ☐ = ☐

d ☐ ÷ ☐ = ☐

e ☐ ÷ ☐ = ☐

f ☐ ÷ ☐ = ☐

g ☐ ÷ ☐ = ☐

h ☐ ÷ ☐ = ☐

i ☐ ÷ ☐ = ☐

j ☐ ÷ ☐ = ☐

Now you know your 3 times tables!

5 Complete the division facts.

a 20 ÷ 5 = ☐

b 18 ÷ 2 = ☐

c 60 ÷ 10 = ☐

d 35 ÷ 5 = ☐

e 14 ÷ 2 = ☐

f 90 ÷ 10 = ☐

6 Now, write a matching multiplication fact.

a ☐ × ☐ = ☐

b ☐ × ☐ = ☐

c ☐ × ☐ = ☐

d ☐ × ☐ = ☐

e ☐ × ☐ = ☐

f ☐ × ☐ = ☐

OXFORD UNIVERSITY PRESS

1 There are 5 chocolates in each box.

How many in:

a 3 boxes? ☐

b 6 boxes? ☐

c 7 boxes? ☐

d 10 boxes? ☐

2 Lindy made 24 cookies. How many will go in each box if she has:

a 3 boxes? ☐

b 6 boxes? ☐

c 8 boxes? ☐

d 2 boxes? ☐

3 The table below shows the number and cost of each item sold at the school fair.

a Complete the table to show how much money each child raised.

b Who sold the most items?

_____

c Who raised the most money?

_____

d How much money would Serena have raised if she sold 8 items?

_____

Name	Number of items sold	Cost per item	Amount raised
Mika	8	$5	
Andy	10	$2	
Serena	6	$10	
Sophia	5	$9	
Hao	9	$4	

e How much money would Mika have raised if he sold 20 items?

_____

f How many items would Sophia have sold if she raised $63? _____

Skip counting can help you to multiply numbers in your head.

The times sign means the same as "groups of".

4×5 is   5,   10,   15,   20

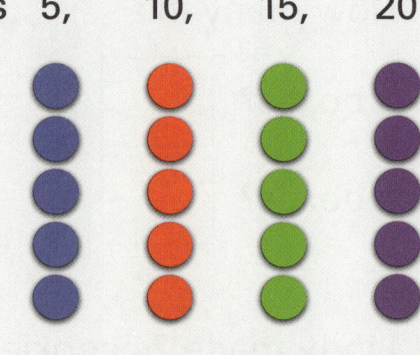

Guided practice

1 Use skip counting to solve.

a 6×3 is   3,   6, _____, _____, _____, _____

b 8×2 is   2,   4, _____, _____, _____, _____, _____, _____

   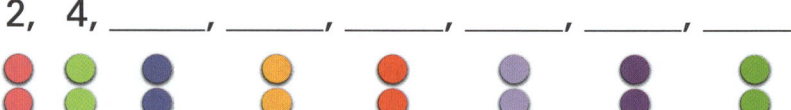

c 3×10 is   _____, _____, _____

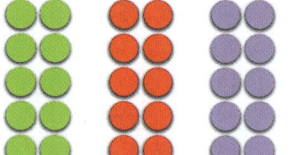   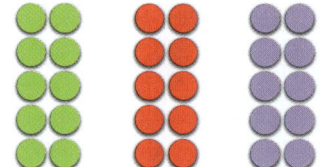

d 7×5 is   _____, _____, _____, _____, _____, _____, _____

   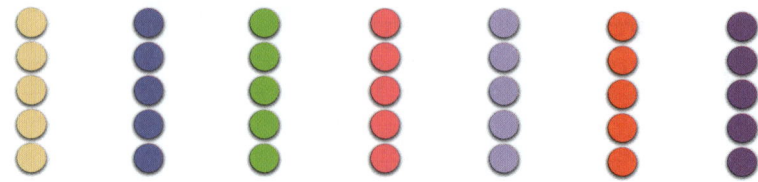

e 8×3 is   _____, _____, _____, _____, _____, _____, _____, _____

   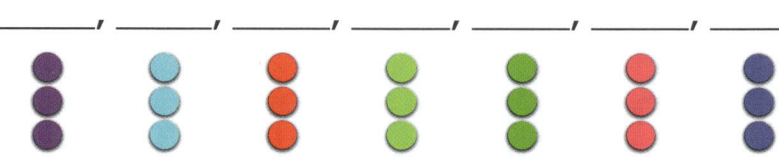

OXFORD UNIVERSITY PRESS

1 To multiply by 4: double, then double again.

$7 \times 4 = 7 \times 2 \times 2 = \quad 14 \times 2 \quad = \quad 28$

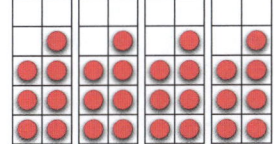 = 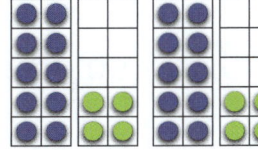 =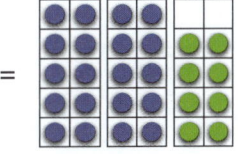

Use double, then double again to solve these sums.

a $8 \times 4 = 8 \times 2 \times 2 = \boxed{} \times 2 = \boxed{}$

b $20 \times 4 = 20 \times 2 \times 2 = \boxed{} \times 2 = \boxed{}$

c $12 \times 4 = \boxed{} \times \boxed{} \times \boxed{} = \boxed{} \times \boxed{} = \boxed{}$

d $30 \times 4 = \boxed{} \times \boxed{} \times \boxed{} = \boxed{} \times \boxed{} = \boxed{}$

To divide by 4: halve, then halve again.

$24 \div 4$ 〈 Halve $\quad 24 \div 2 = 12$
Halve again $\quad 12 \div 2 = 6$ $\qquad$ So $24 \div 4 = 6.$

2 Use halve, then halve again to solve.

a $16 \div 4$ 〈 Halve $\quad 16 \div 2 = \boxed{}$
Halve again $\quad \boxed{} \div 2 = \boxed{}$ $\qquad$ So $16 \div 4 = \boxed{}$.

b $40 \div 4$ 〈 Halve $\quad 40 \div 2 = \boxed{}$
Halve again $\quad \boxed{} \div 2 = \boxed{}$ $\qquad$ So $40 \div 4 = \boxed{}$.

c $60 \div 4$ 〈 Halve $\quad 60 \div 2 = \boxed{}$
Halve again $\quad \boxed{} \div 2 = \boxed{}$ $\qquad$ So $60 \div 4 = \boxed{}$.

Multiplication facts can help with division.

15 ÷ 3   Think   3 × ⬚?⬚ = 15.   The answer is 5.

3   Solve these sums.

a   26 ÷ 2   Think   2 × ⬚13⬚ = 26,   so 26 ÷ 2 = ⬚ .

b   27 ÷ 3   Think   3 × ⬚ = 27,   so 27 ÷ 3 = ⬚ .

c   45 ÷ 5   Think   5 × ⬚ = 45,   so 45 ÷ 5 = ⬚ .

d   55 ÷ 5   Think   5 × ⬚ = 55,   so 55 ÷ 5 = ⬚ .

e   120 ÷ 10   Think   10 × ⬚ = 120, so 120 ÷ 10 = ⬚ .

4   Solve using known facts.

Do you know any other shortcuts to help you work out multiplication and division in your head?

a   How many chocolates in 5 packets of 6?   ⬚

b   How many pencils in 10 packets of 9?   ⬚

c   How many cookies go in each bag if you have 60 cookies and 6 bags?   ⬚

d   How many cookies go in each bag if you have 24 cookies and 8 bags?   ⬚

e   How many people in 4 rows of 8?   ⬚

f   If 36 people get on a plane, how many rows of 3 can they fill?   ⬚

g   How many rows of 6 can they fill?   ⬚

h   How much money would you earn if you were paid $8 an hour for 10 hours?   ⬚

OXFORD UNIVERSITY PRESS

Extended practice

1 Use your choice of strategy to solve.

a Four teams with 16 people in each were going to the stadium. How many seats were needed on the bus?

b At the end of the game 84 people were divided equally onto 4 buses. How many people on each bus?

c The front section of the stadium has 5 rows with 12 seats in each. How many people can sit there?

d 200 oranges were shared between 10 teams. How many oranges did each team get?

OXFORD UNIVERSITY PRESS

You can split larger numbers to make multiplying easier.

3×17 is the same as 3×10 + 3×7 = $30 + 21$

= 51

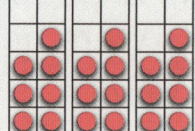

You can also use the split strategy to help multiply numbers in your head.

Guided practice

1 Use the split strategy to solve these sums.

a 2×26 is the same as $2 \times \boxed{}$ + $2 \times \boxed{}$ = $\boxed{}$ + $\boxed{}$

= $\boxed{}$

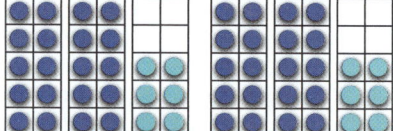
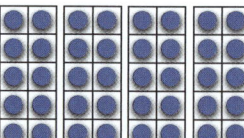
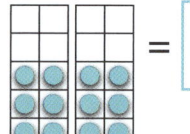

b 4×14 is the same as $4 \times \boxed{}$ + $4 \times \boxed{}$ = $\boxed{}$ + $\boxed{}$

= $\boxed{}$

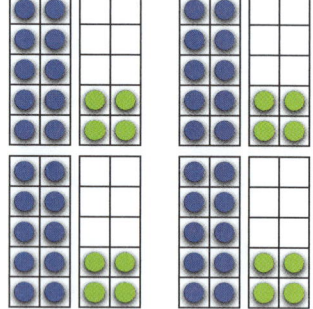

c 3×19 is the same as $3 \times \boxed{}$ + $3 \times \boxed{}$ = $\boxed{}$ + $\boxed{}$

= $\boxed{}$

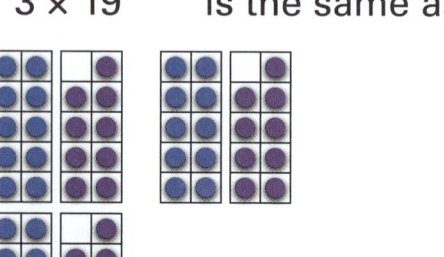
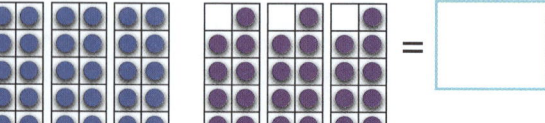

OXFORD UNIVERSITY PRESS

1 Solve with the split strategy.

a $5 \times 13 = 5 \times \boxed{} + 5 \times \boxed{} = \boxed{} + \boxed{}$

$\phantom{5 \times 13 = 5 \times \boxed{00} + 5 \times \boxed{00} = } = \boxed{}$

b $6 \times 21 = 6 \times \boxed{} + 6 \times \boxed{} = \boxed{} + \boxed{}$

$\phantom{6 \times 21 = 6 \times \boxed{00} + 6 \times \boxed{00} = } = \boxed{}$

c $4 \times 32 = 4 \times \boxed{} + 4 \times \boxed{} = \boxed{} + \boxed{}$

$\phantom{4 \times 32 = 4 \times \boxed{00} + 4 \times \boxed{00} = } = \boxed{}$

d $7 \times 24 = \boxed{} \times \boxed{} + \boxed{} \times \boxed{} = \boxed{} + \boxed{}$

$\phantom{7 \times 24 = \boxed{00} \times \boxed{00} + \boxed{00} \times \boxed{00} = } = \boxed{}$

e $5 \times 45 = \boxed{} \times \boxed{} + \boxed{} \times \boxed{} = \boxed{} + \boxed{}$

$\phantom{5 \times 45 = \boxed{00} \times \boxed{00} + \boxed{00} \times \boxed{00} = } = \boxed{}$

f $8 \times 33 = \boxed{} \times \boxed{} + \boxed{} \times \boxed{} = \boxed{} + \boxed{}$

$\phantom{8 \times 33 = \boxed{00} \times \boxed{00} + \boxed{00} \times \boxed{00} = } = \boxed{}$

g $3 \times 58 = \boxed{} \times \boxed{} + \boxed{} \times \boxed{} = \boxed{} + \boxed{}$

$\phantom{3 \times 58 = \boxed{00} \times \boxed{00} + \boxed{00} \times \boxed{00} = } = \boxed{}$

You can also use a grid for the split strategy.

$6 \times 23 =$

×	20	3
6	120	18

$= \boxed{138}$

Add the two answers at the bottom of the grid to find the total.

2 Solve with the grid method.

a $4 \times 27 =$

×	20	7
4		

$=$

b $6 \times 36 =$

×	30	6
6		

$=$

c $5 \times 53 =$

×		
5		

$=$

d $3 \times 62 =$

×		
3		

$=$

e $5 \times 84 =$

×		
5		

$=$

f $4 \times 48 =$

×		
4		

$=$

g $2 \times 95 =$

×		
2		

$=$

OXFORD UNIVERSITY PRESS

1 Solve using your choice of written methods. Show how you got your answer.

a

> 4 × 37

b

> 6 groups of 16

c Morgan bought 5 sets of basketball cards with 38 in each pack. How many cards does he have?

d Nouf ordered 1 doughnut for each of her birthday guests and 3 extras, in case more guests arrived. She bought 4 boxes with 26 doughnuts in each. How many guests was she expecting?

It's easy to make friends with addition and multiplication.

You choose how to start and the answer is the same.

2 + 3 or 3 + 2

 = 5

3 x 2   or   2 x 3

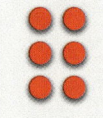

3 x 2 = 6   2 x 3 = 6

You can group the numbers in any way.

4 + 2 + 3 = ?

4   +   5   = 9

6   +   3 = 9

4 x 3 x 2 = ?

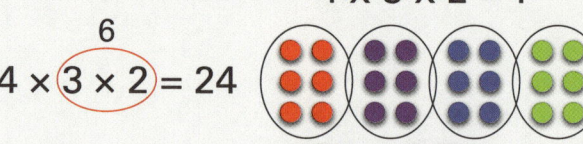

6
4 × (3 × 2) = 24

12
(4 × 3) × 2 = 24

What would happen with subtraction and division?

Guided practice

1 Find the answers in two ways.

a 13 + 5 = ☐   and   5 + 13 = ☐

b 15 x 2 = ☐   and   2 x 15

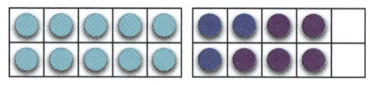   

c (4 + 7) + 3 = ☐   and   4 + (7 + 3) = ☐

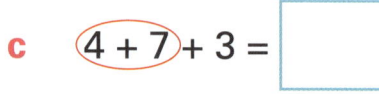   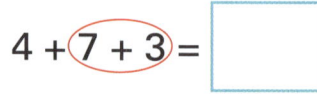

d (4 x 5) x 2 = ☐   and   4 x (5 x 2) = ☐

OXFORD UNIVERSITY PRESS

Is one way easier than the other?

1 Find the answers in two ways.

a   23 + 5 =   [ ]   and   5 + 23 =   [ ]

b   14 + 24 =   [ ]   and   24 + 14 =   [ ]

c   8 + 2 + 16 =   [ ]   and   16 + 2 + 8 =   [ ]

d   3 + 12 + 7 =   [ ]   and   7 + 3 + 12 =   [ ]

2 Change the order to find an easy way to add.

e.g. What is 7 + 9 + 3 + 1?   10 + 10   7 + 3 + 9 + 1 = 20

a   What is 6 + 7 + 4 + 3?   _____ =   [ ]

b   What is 18 + 5 + 2 + 5?   _____ =   [ ]

c   What is 14 + 9 + 6 + 1?   _____ =   [ ]

d   What is 23 + 6 + 14 + 7?   _____ =   [ ]

3 Change the order to find an easy way to multiply.

e.g. What is $6 \times 2 \times 5$?   = 10 $2 \times 5 \times 6 = 10 \times 6 = 60$

a   What is $5 \times 7 \times 2$?   _____ =   [ ]

b   What is $6 \times 2 \times 3$?   _____ =   [ ]

c   What is $3 \times 5 \times 2$?   _____ =   [ ]

d   What is $2 \times 7 \times 3$?   _____ =   [ ]

Addition and subtraction are linked. Multiplication and division are linked, too. Knowing this is a good way to check your work.

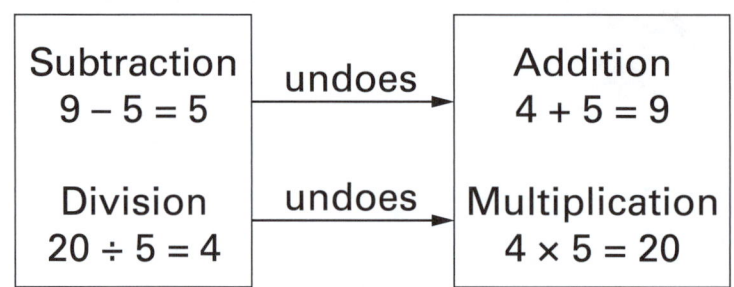

Subtraction	undoes	Addition
$9 - 5 = 5$	→	$4 + 5 = 9$
Division	undoes	Multiplication
$20 \div 5 = 4$	→	$4 \times 5 = 20$

4 Find the answers. Check by "undoing" the problem.

a $14 + 9 =$ ☐   Check ☐ − ☐ = ☐

b $25 - 14 =$ ☐   Check ☐ + ☐ = ☐

c $9 \times 3 =$ ☐   Check ☐ ÷ ☐ = ☐

d $40 \div 5 =$ ☐   Check ☐ × ☐ = ☐

e $42 - 21 =$ ☐   Check ☐ + ☐ = ☐

f $11 \times 5 =$ ☐   Check ☐ ÷ ☐ = ☐

g $43 + 24 =$ ☐   Check ☐ − ☐ = ☐

h $40 \div 5 =$ ☐   Check ☐ × ☐ = ☐

5 Look for shortcuts to solve the problems. Be ready to explain how you get your answers.

a $3 + 3 + 3 + 3 + 3 =$ ☐   b $3 + 4 + 17 =$ ☐

c $2 \times 9 \times 5 =$ ☐   d $18 + 7 + 3 + 12 =$ ☐

e $4 + 4 + 4 + 4 + 4 + 4 =$ ☐   f $90 \div 10 =$ ☐

g $3 + 16 + 8 + 7 + 2 + 14 =$ ☐   h $7 + 7 + 7 + 7 + 8 =$ ☐

OXFORD UNIVERSITY PRESS

1

a Tran's football card book has 15 pages. There are 10 cards on each page. Jack's book has 10 pages with 15 cards on each page. Tran thinks he has more cards than Jack. Is Tran right? How many cards does each person have?

b Eva got pocket money for doing some jobs. The table shows how much she got over 10 weeks. How much did Eva get altogether?

Week	1	2	3	4	5	6	7	8	9	10
Amount	$3	$8	$4	$7	$12	$11	$5	$9	$5	$16

c Jalia read the following pages in a week:

Monday: 9 pages, Tuesday: 9 pages, Wednesday: 9 pages, Thursday: 9 pages, Friday: 9 pages, Saturday: 9 pages, Sunday: 10 pages.

How many pages did Jalia read altogether?

d Henry's grandmother has six shelves of books. She wants to share them between her five grandchildren. She counts this many books on each shelf: 13 books, 18 books, 24 books, 17 books, 22 books, 16 books. How many books did each grandchild receive?

The **numerator** tells us how many parts we are dealing with.

The **denominator** tells us how many parts a whole or group is divided into.

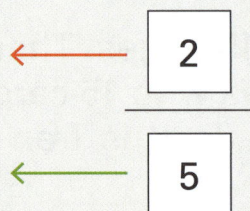

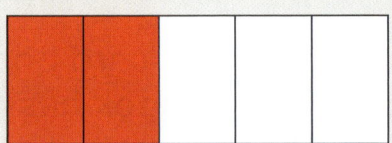

Two-fifths or **2** parts out of **5** are shaded.

The numerator is the top number of the fraction. The denominator is the bottom number of the fraction.

Guided practice

1 Shade the fractions.

a $\dfrac{3}{5}$

b $\dfrac{1}{3}$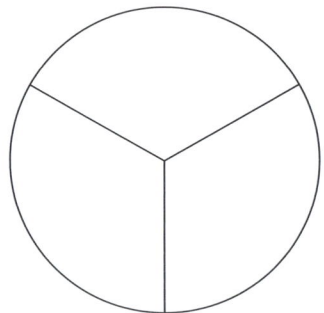

c $\dfrac{1}{2}$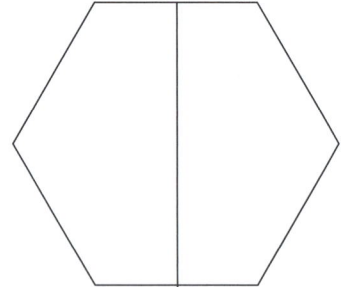

d $\dfrac{3}{4}$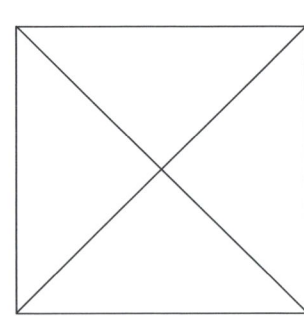

e $\dfrac{4}{5}$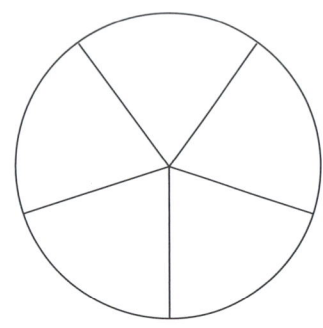

f $\dfrac{2}{3}$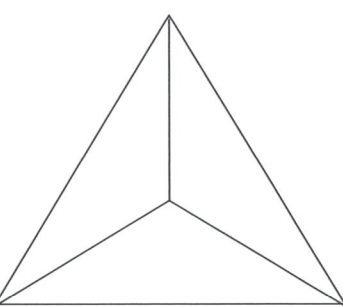

OXFORD UNIVERSITY PRESS

Independent practice

1 What fraction is shaded?

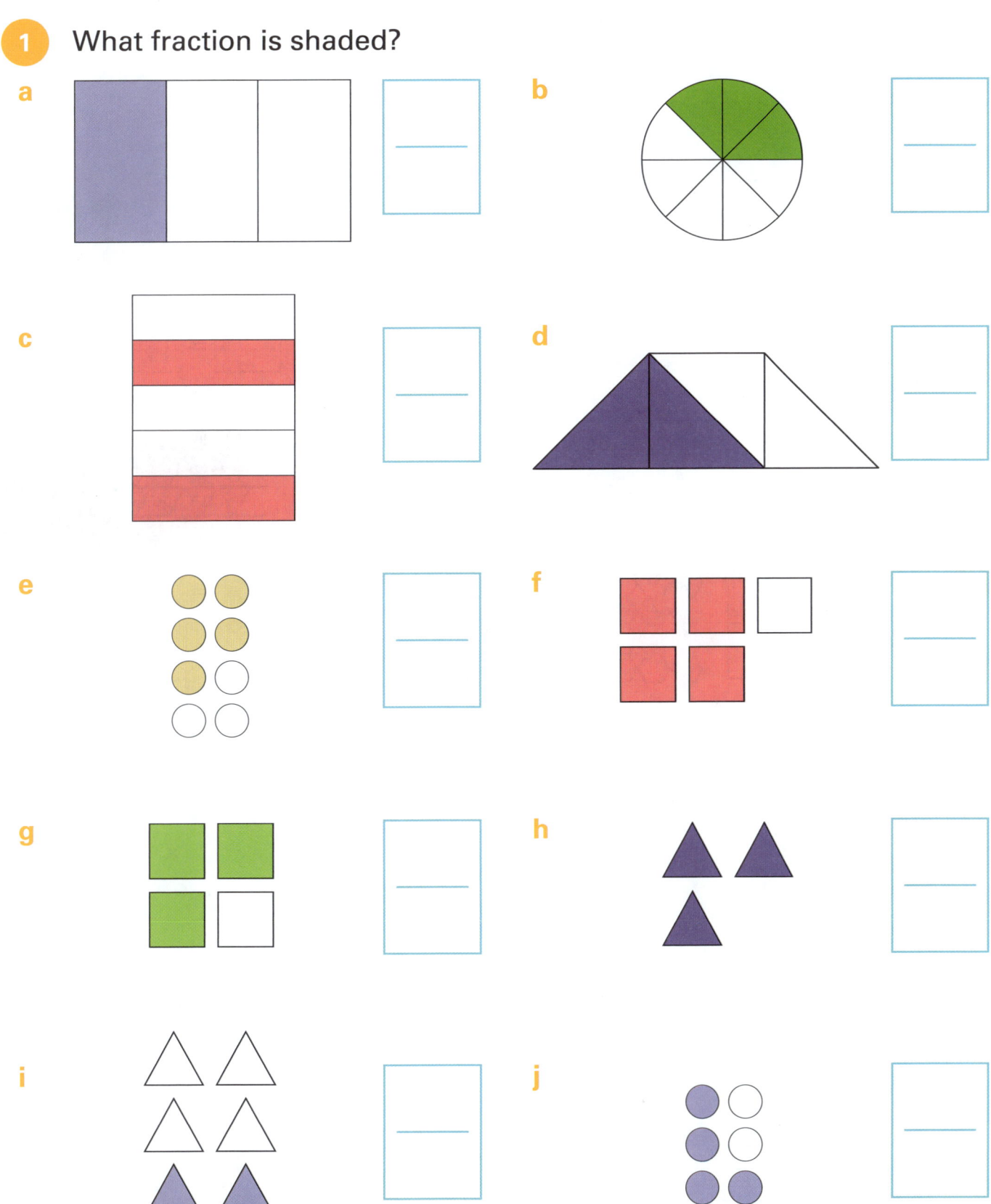

a

b

c

d

e

f

g

h

i

j

2 Draw lines to match each fraction with its picture.

| $\dfrac{1}{2}$ | $\dfrac{1}{3}$ | $\dfrac{3}{4}$ | $\dfrac{2}{8}$ | $\dfrac{3}{5}$ | $\dfrac{2}{5}$ |

> Remember that the parts of a fraction need to be equal in size.

3 Divide each rectangle into the fraction shown.

a

b

c

d

quarters   fifths   thirds   halves

4 Which fraction in question 3 has:

a   the most parts?   _____

b   the least parts?   _____

c   the smallest parts?   _____

d   the biggest parts?   _____

OXFORD UNIVERSITY PRESS

Extended practice

1

a Draw a line to divide the square into 2 equal parts.

b What fraction is each part?

c Draw another line to make 4 equal parts.

d What fraction is each part?

e Draw 2 more lines to make 8 equal parts.

f What fraction is each part?

g Colour in 5 parts.

h What fraction is coloured in?

i What fraction is **not** coloured in?

2 Order the fractions from smallest to largest.

$$\frac{1}{2} \qquad \frac{1}{8} \qquad \frac{1}{4} \qquad \frac{5}{8}$$

_____ _____ _____ _____

Fractions on number lines

Number lines are useful for counting by and comparing fractions.

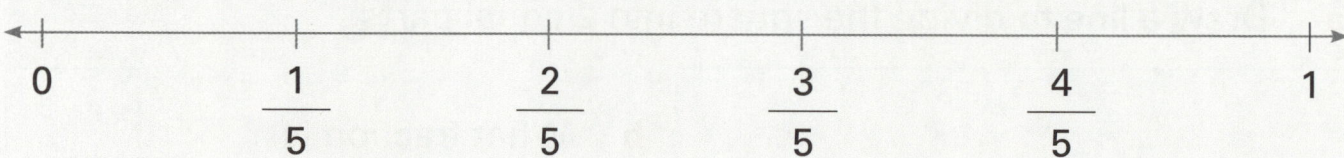

What is another way we could write 1 on this number line?

1 Fill in the missing fractions.

a

b

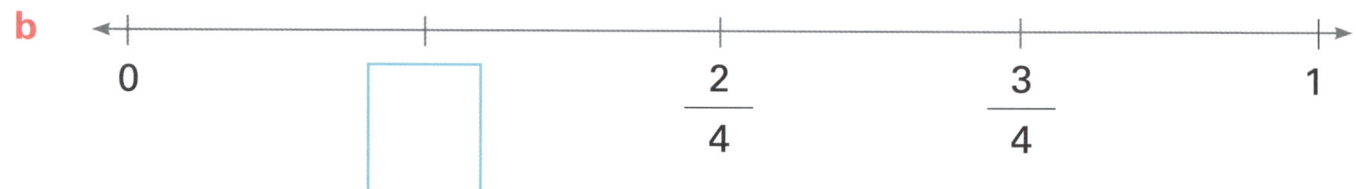

c

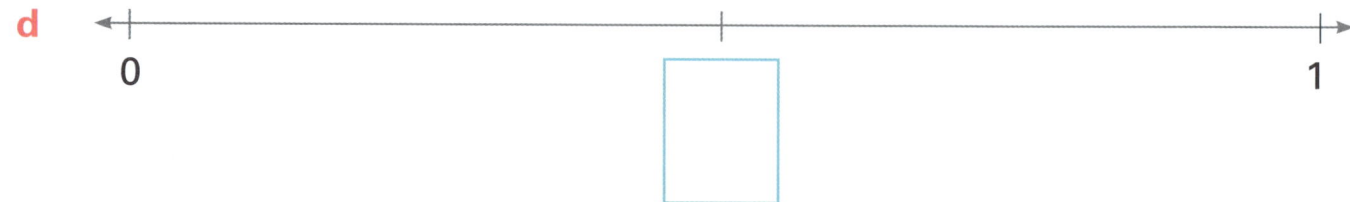

d

OXFORD UNIVERSITY PRESS

1 Match the fractions to the correct place on the number line.

a

| $\frac{4}{4}$ | $\frac{2}{4}$ | $\frac{1}{4}$ | $\frac{3}{4}$ |

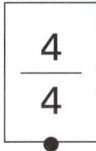

0

b

| $\frac{2}{5}$ | $\frac{4}{5}$ | $\frac{5}{5}$ | $\frac{1}{5}$ | $\frac{3}{5}$ |

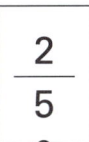

0

c

| $\frac{4}{8}$ | $\frac{7}{8}$ | $\frac{8}{8}$ | $\frac{2}{8}$ | $\frac{6}{8}$ | $\frac{1}{8}$ | $\frac{5}{8}$ |

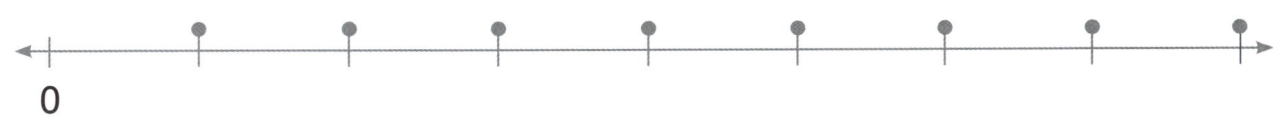

0

2 Which fraction is missing from question 1c? _____

3 How many:

a eighths in 1? ⬚

b halves in 1? ⬚

c fifths in 1? ⬚

d thirds in 1? ⬚

e quarters in 1? ⬚

4 Use the number lines to decide which fraction is bigger.

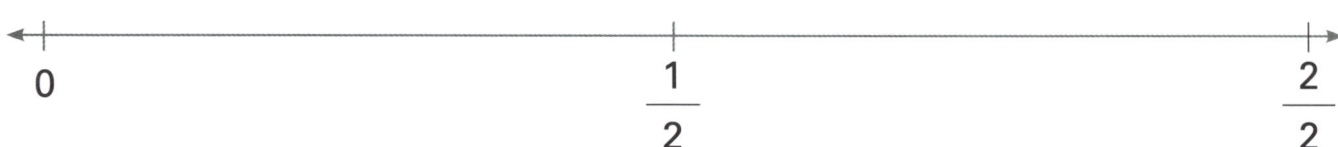

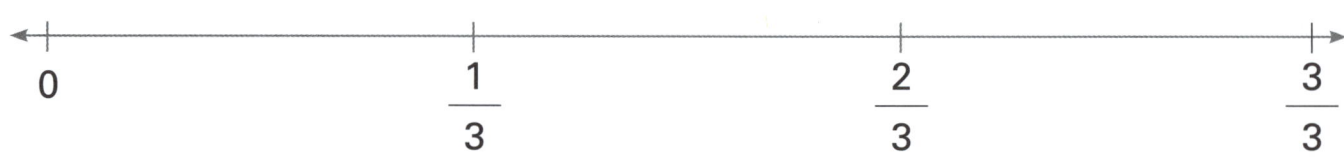

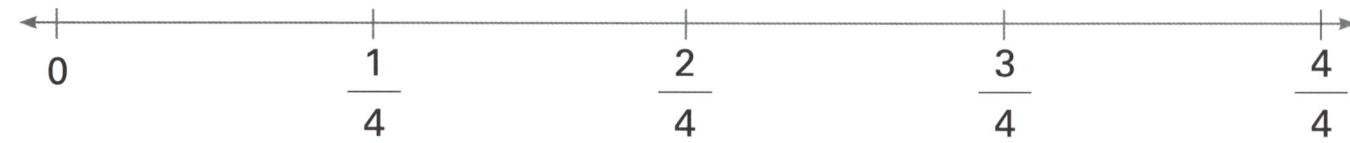

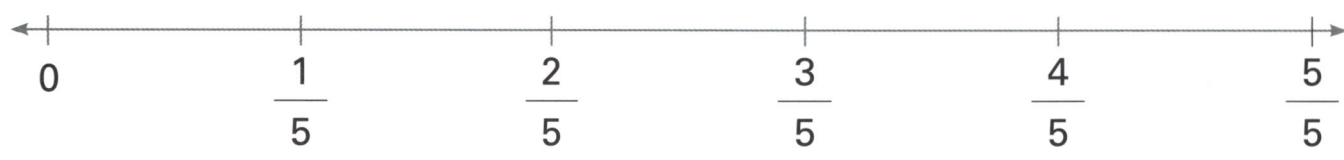

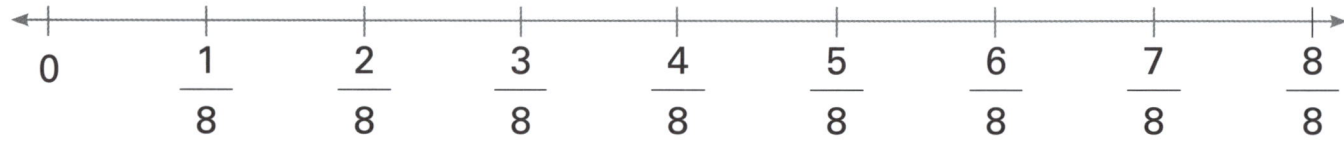

a $\dfrac{1}{2}$ or $\dfrac{1}{4}$? _____

b $\dfrac{1}{5}$ or $\dfrac{1}{8}$? _____

c $\dfrac{1}{5}$ or $\dfrac{1}{3}$? _____

d $\dfrac{3}{8}$ or $\dfrac{2}{4}$? _____

e $\dfrac{2}{3}$ or $\dfrac{2}{5}$? _____

f $\dfrac{4}{8}$ or $\dfrac{4}{5}$? _____

5 Explain why $\dfrac{5}{5}$ and $\dfrac{8}{8}$ are the same size.

Which other fractions are the same size as $\frac{1}{2}$?

_____

_____

_____

OXFORD UNIVERSITY PRESS

You can also count by fractions beyond 1.

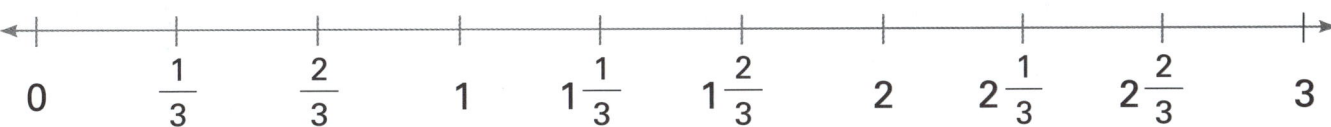

1 Fill in the missing fractions.

a

0    $\frac{1}{4}$ ☐   ☐   1 $1\frac{1}{4}$ ☐   ☐   ☐

b

0   ☐   ☐ $\frac{3}{5}$ $\frac{4}{5}$ ☐   ☐ $1\frac{2}{5}$ ☐   ☐   2

c

0    $\frac{1}{8}$ ☐ $\frac{3}{8}$ ☐ ☐ ☐ $\frac{7}{8}$ ☐ $1\frac{1}{8}$ ☐ ☐ ☐ $1\frac{5}{8}$ ☐ ☐   2

2

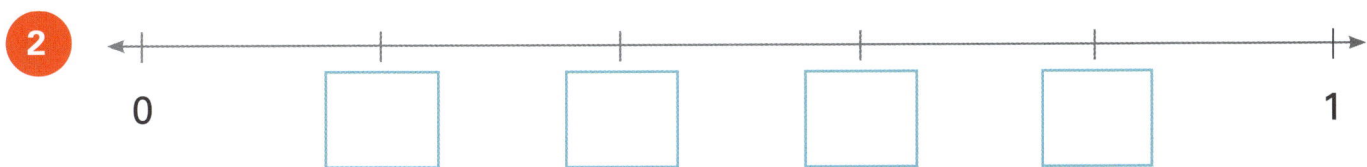

0   ☐   ☐   ☐   ☐   1

a How many segments has the number line been divided into? ☐

b What is each segment called? _____

c Write in the missing fractions.

d What would the next number on the number line be? _____

e How could you rename 1 on the number line as a fraction? _____

You can make 50c in different ways using these coins.

 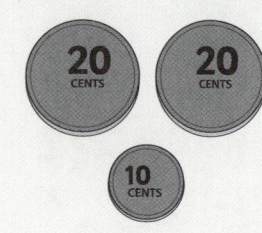  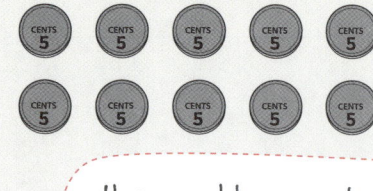

How could you make 50c using 4 coins?

Guided practice

1 Using the coins from above, show 3 ways to make these amounts.

a 70c

b $1

c 40c

OXFORD UNIVERSITY PRESS

Independent practice

1 Using the coins we have looked at in this topic, draw 3 coins to make these amounts.

a 30c

b 90c

c $1.20

d $2.10

2 Using the coins we have looked at in this topic, show the smallest number of coins you could use to buy these items.

a

b

c

d

e

f

3 How much change would you get from $5?

a

Price $3.50

b

Price $1.75

c

Price 90c

d

Price $2.05

A good way to calculate change is to count up from the amount the item costs to the amount you are paying.

4 Show the change amounts.

a

$3.20

b

$1.10

c

$5.65

d

$1.60

$1.60

OXFORD UNIVERSITY PRESS

Extended practice

1 Money amounts not ending in 0 or 5c are sometimes rounded. Round these amounts to the nearest 5c.

a 21c _____ **b** 68c _____ **c** 44c _____

d $1.03 _____ **e** $1.78 _____ **f** $2.99 _____

2

a Count how much money Florcita has.

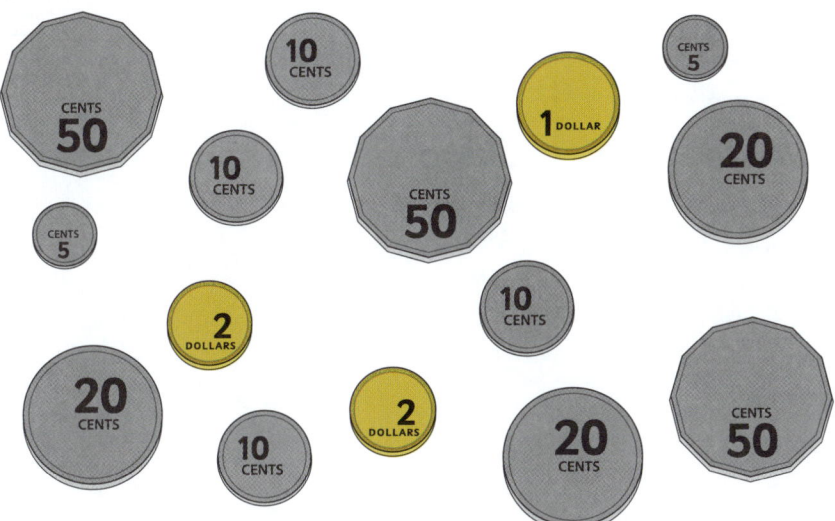

b List 2 different coin combinations she could use to buy a toy that costs $5.30.

c How much change would Florcita get if she bought a toy that cost $7.58? Explain the reason for your answer.

_____

_____

Rule: Add 3

| 2 | 5 | 8 | 11 | 14 | 17 | 20 | 23 | 26 | 29 |

> Each number in the pattern is 3 bigger than the one before it.

Guided practice

1 Follow the rule to finish the pattern.

a Rule: Add 5

| 3 | 8 | 13 | | | | | | | |

b Rule: Subtract 3

| 54 | 51 | 48 | | | | | | | |

c Rule: Add 6

| 6 | 12 | | | | | | | | |

d Rule: Subtract 4

| 65 | 61 | 57 | | | | | | | |

e Rule: Add 10

| 24 | 34 | 44 | | | | | | | |

OXFORD UNIVERSITY PRESS

Independent practice

1 Write a rule for the number patterns.

a Rule: _____

| 3 | 13 | 23 | 33 | 43 | 53 | 63 | 73 |

b Rule: _____

| 90 | 85 | 80 | 75 | 70 | 65 | 60 | 55 |

c Rule: _____

| 4 | 11 | 18 | 25 | 32 | 39 | 46 | 53 |

2 Fill in the missing numbers and write the rule.

a

In	Out
52	48
36	32
44	
28	

Rule: _____

b

In	Out
13	11
31	29
5	
47	

Rule: _____

c

In	Out
19	27
44	52
62	
53	

Rule: _____

d

In	Out
64	55
48	39
56	
30	

Rule: _____

3

a Complete the diagram and number pattern.

●	●●●	●●●●	●●●●●		
1	3				

b What is the rule? _____

4

a Complete the diagram and number pattern.

●●●●●●●●●●●●●●●●●●	●●●●●●●●●●●●●●●	●●●●●●●●●●●●			
18	15				

b What is the rule? _____

The numbers in addition patterns get bigger and the numbers in subtraction patterns get smaller.

5

a Make your own addition pattern.

Rule: _____

b Make your own subtraction pattern.

Rule: _____

OXFORD UNIVERSITY PRESS

This pattern has a 2-step rule.

Rule: Add 4, subtract 2

0	4	2	6	4	8	6	10	8	12	10

1 Write the 2-step rules.

a Rule: _____

0	5	4	9	8	13	12	17	16	21	20

b Rule: _____

20	18	21	19	22	20	23	21	24	22	25

2 Follow the rule to finish the pattern.

a Rule: Add 1, add 3

1	2	5	6						

b Rule: Subtract 2, subtract 3

56	54	51	49						

3 Make your own 2-step pattern.

Rule: _____

Missing numbers

The = sign shows that both sides are the same.

8 + ☐ = 11

 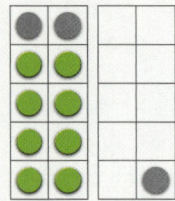

We need 3 more to make 11, so the missing number is 3.

How could you use subtraction to solve this problem?

Guided practice

1 Use the ten frames to find the missing numbers.

a 7 + ☐ = 12

b 19 − ☐ = 15

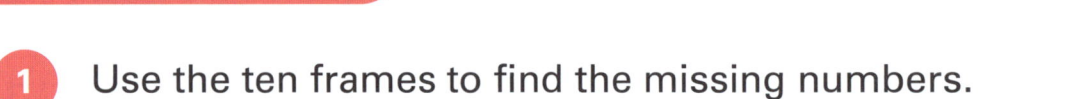

c 10 + ☐ = 18

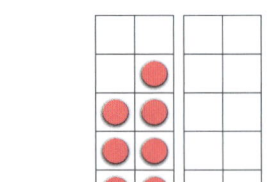 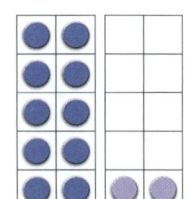

d 16 − ☐ = 9

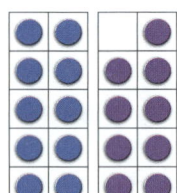 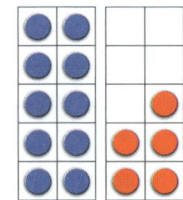

e 17 = ☐ + 14

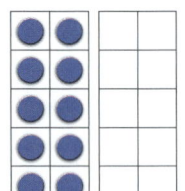 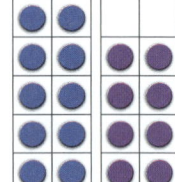

f 16 = 19 − ☐

 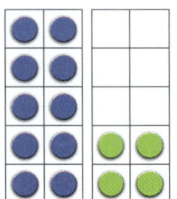 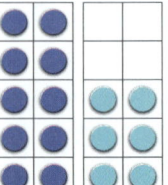 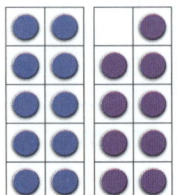

OXFORD UNIVERSITY PRESS

Independent practice

1 Make the equations balance.

a $10 + 2 = 8 + \boxed{}$

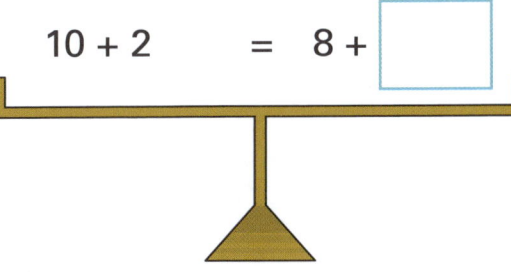

b $13 - 5 = 10 - \boxed{}$

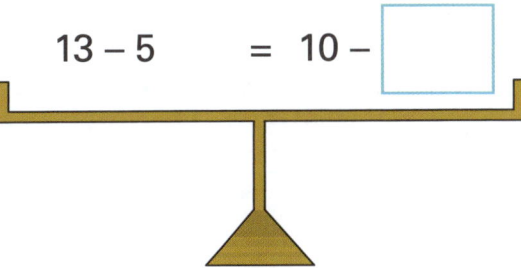

c $13 + 4 = \boxed{} + 9$

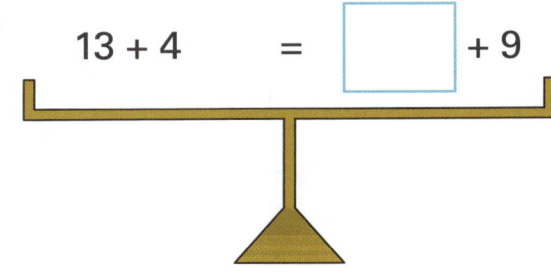

d $33 - 3 = 15 + \boxed{}$

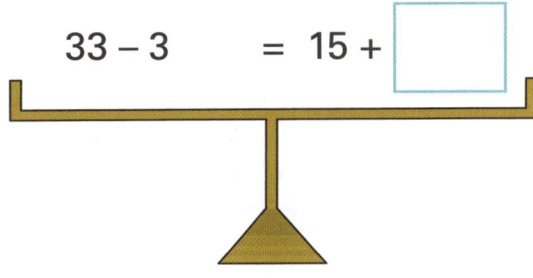

e $28 + 2 = \boxed{} + 10$

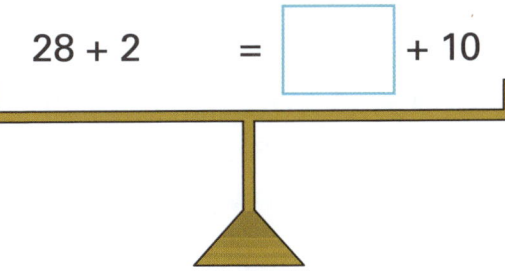

f $17 + 11 = 8 + \boxed{}$

2 Use + or – to complete.

a $8 \boxed{} 6 = 14$

b $11 \boxed{} 7 = 18$

c $11 \boxed{} 7 = 4$

d $24 \boxed{} 12 = 36$

e $34 \boxed{} 8 = 26$

f $19 \boxed{} 9 = 10$

g $45 \boxed{} 20 = 25$

h $25 \boxed{} 20 = 45$

3 Write a number sentence to solve the word problems.

a Anjali sold 46 cakes on Saturday and 19 on Sunday. How many did she sell on the weekend?

b Marco made 84 wind chimes. He sold 32 at the market. How many are left?

How do you know whether to use addition or subtraction?

c Kristy earned $74 and Felix earned $49. How much more did Kristy earn?

d Spiro read 42 pages of his book on Monday, 14 on Tuesday and 28 on Wednesday. How many pages did he read altogether?

e Gordana needs 200 points to get to the next level. She has 153. How many more points does she need?

f 100 eggs were delivered to the bakery. The baker used 32 on Monday and 41 on Tuesday. How many are left?

OXFORD UNIVERSITY PRESS

Extended practice

1 Class 3M are tracking how many steps they take in a day. The table shows the steps taken by one group in one hour.

Name	Steps
Jonas	97
Sumi	131
Megan	164
George	46
Tanmay	253
Daina	98

a How many steps did Sumi and Megan take altogether?

b George and which other student's total is 144?

c How many more steps did Sumi take than Jonas?

d Which 2 students' steps total 350?

e Use a calculator to find the total steps the 6 students took.

f How many more steps did Megan and Sumi take than Tanmay?

2 True or false?

a 23 + 32 = 60 − 7 | True | False |

b 50 − 29 = 8 + 13 | True | False |

c 26 − 15 = 37 − 26 | True | False |

d 58 + 24 = 99 − 17 | True | False |

e 48 + 52 = 9 + 91 | True | False |

Length

Shorter lengths are measured in centimetres (cm).

Longer lengths are measured in metres (m). There are 100 cm in 1 m.

The length of the eraser is 4 cm.

0 CM 1 2 3 4 5 6

In real life, the guitar would be 96 cm longer than the eraser.

0 10 20 30 40 50 60 70 80 90 100

The length of the guitar is 100 cm or 1 m.

Guided practice

1 Use a ruler to find the lengths of these items.

a [ ] cm

b [ ] cm

c [ ] cm

d [ ] cm

2

a Which item is longest? _____

b Which item is shortest? _____

c Which item is 5 cm long? _____

OXFORD UNIVERSITY PRESS

Independent practice

1 **a** Choose an item in the classroom that you think matches each length listed in the table. Record the item in the table.

b Now measure the items and record the actual lengths.

Length	Item	Actual length
10 cm		
30 cm		
50 cm		
1 m		
3 m		
1 m 50 cm		

2 Would you use cm or m to measure the length of:

a the classroom?   **b** this book?

c a basketball court?   **d** your house?

e a chocolate bar?   **f** a glue stick?

3 Circle the best estimate for the length of:

a a smart phone.   30 cm   13 cm   13 m

b a car.   5 m   5 cm   50 m

c a pet turtle.   18 m   10 m   12 cm

d an elephant.   6 cm   60 cm   6 m

Area

A square centimetre is 1 cm wide and 1 cm high.

We use square centimetres to measure area.

The abbreviation of square centimetres is cm².

1 cm

1 cm

What does area mean?

Area = 10 cm²

Guided practice

1 Record the area of each shape.

a 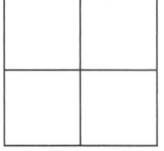 ⬜ cm²

b 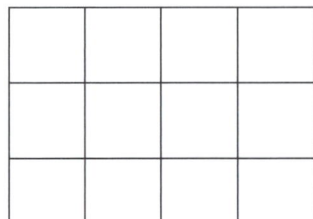 ⬜ cm²

c  ⬜ cm²

d 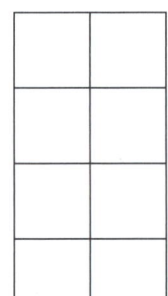 ⬜ cm²

e 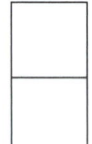 ⬜ cm²

f  ⬜ cm²

2 Write the letter of the shape that has:

a the largest area. _____

b the smallest area. _____

3 Which 2 shapes have the same area? _____

OXFORD UNIVERSITY PRESS

1 Use the cm² grid paper to draw:

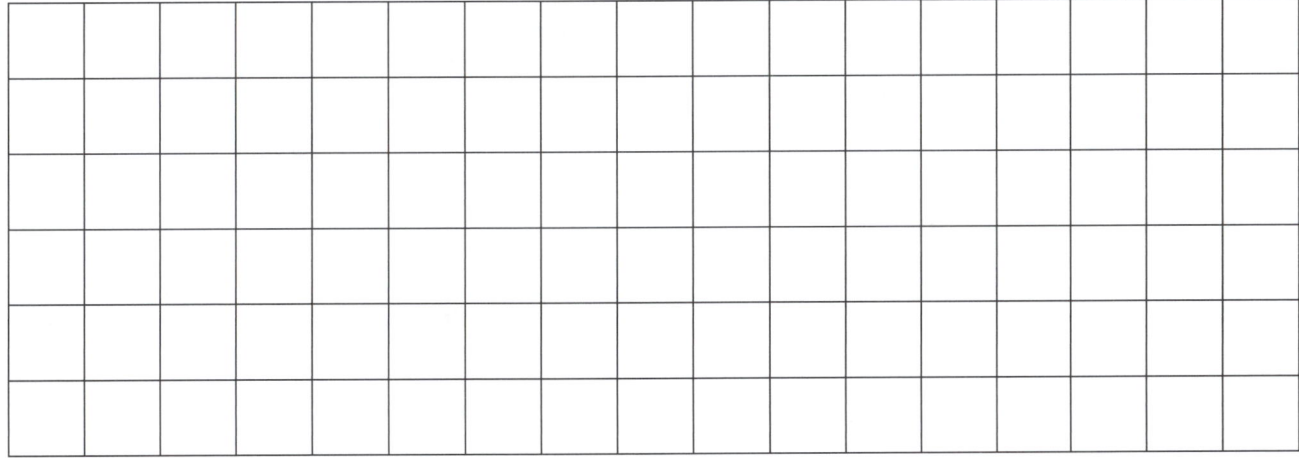

a a blue square with an area of 9 cm².

b a red square with an area of 10 cm².

c 2 different green squares, each with an area of 12 cm².

d a yellow square with an area of 4 cm².

2 What is the total area of the shapes in question 1? _____ cm²

3

a Estimate the area of the shape below. _____ cm²

b Find the area of the blue square. _____ cm²

c Find the area of the red rectangle. _____ cm²

d What is the total area? _____ cm²

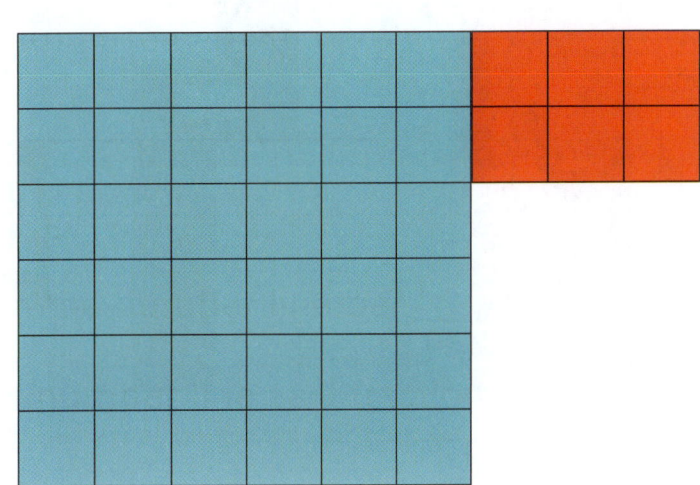

1 Millimetres (mm) are used to measure very small lengths, or when you need very accurate measurements. There are 10 mm in 1 cm.

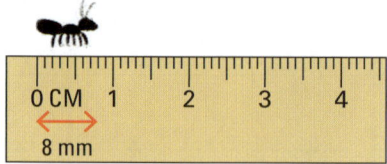

Measure these lines in mm.

a _____ mm   **b** _____ mm

c _____ mm   **d** _____ mm

e _____ mm   **f** _____ mm

2 Square metres are used for measuring large areas. A square metre (m²) is 100 cm by 100 cm.

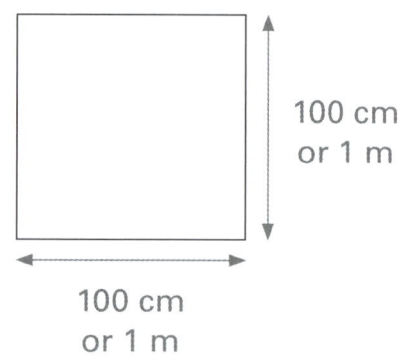

100 cm or 1 m

100 cm or 1 m

PLAN OF MY BACKYARD

10 m

8 m

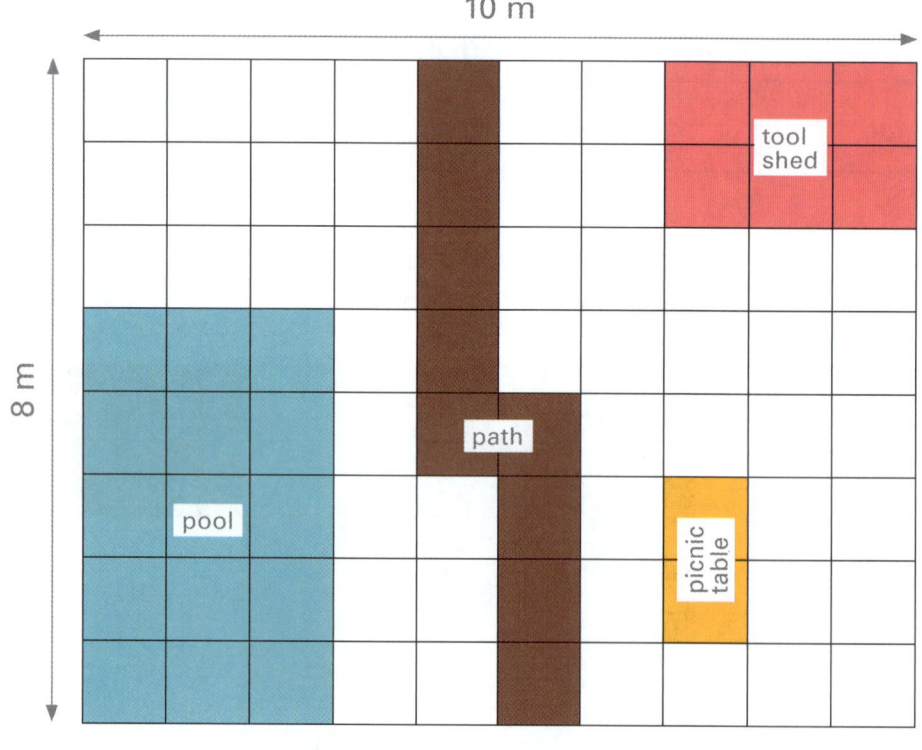

Record the area of:

a the tool shed.

_____ m²

b the pool.

_____ m²

c the picnic table.

_____ m²

d the path.

_____ m²

3 How much bigger is the tool shed than the picnic table? _____ m²

OXFORD UNIVERSITY PRESS

Volume

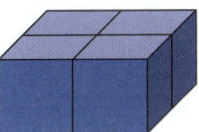

1 cm

1 cm

1 cm

This centicube is 1 cm high, 1 cm wide and 1 cm long. It is also called a cubic centimetre or 1 cm³.

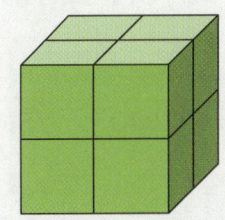

This object has a volume of 8 centicubes or 8 cm³.

What different meanings can the word volume have?

Guided practice

1 Write the volumes.

a

_____ cubic centimetres

or _____ cm³

b

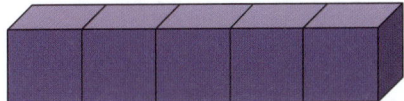

_____ cubic centimetres

or _____ cm³

c

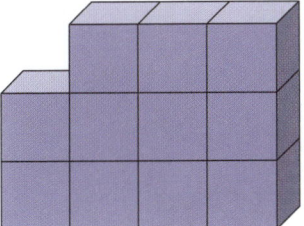

_____ cubic centimetres

or _____ cm³

d

_____ cubic centimetres

or _____ cm³

e

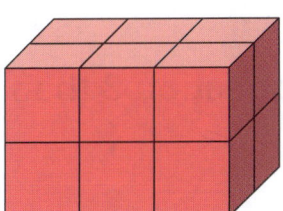

_____ cubic centimetres

or _____ cm³

f

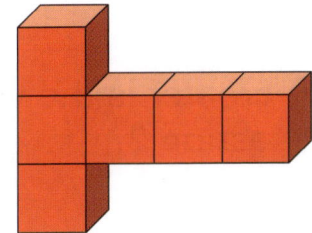

_____ cubic centimetres

or _____ cm³

Use the layers to find the volume.

1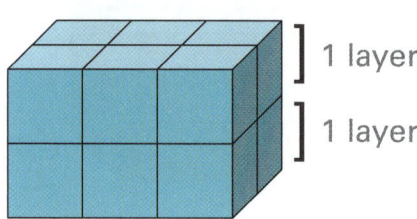
1 layer
1 layer

a How many layers? _____

b How many cubic centimetres in each layer? ___6___

c Total volume: _____ cm³

2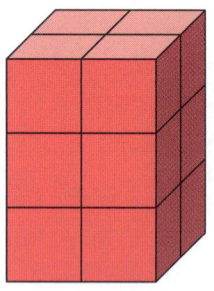

a How many layers? _____

b How many cubic centimetres in each layer? _____

c Total volume: _____ cm³

3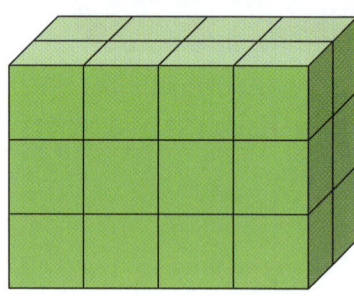

a How many layers? _____

b How many cubic centimetres in each layer? _____

c Total volume: _____ cm³

4

a Name the colour of the object above with the biggest volume.

_____

b Name the colours of the objects with the same volume.

_____   _____

c How much greater is the volume of the object in question 3 than the object in question 2?

_____

Capacity

Millilitres (mL) and litres (L) are two units of capacity.

There are 1000 mL in 1 L.

A soft drink can holds less than 1 L.

375 mL

1 L

2 L

A large paint tin holds more than 1 L.

How is capacity different from volume?

1

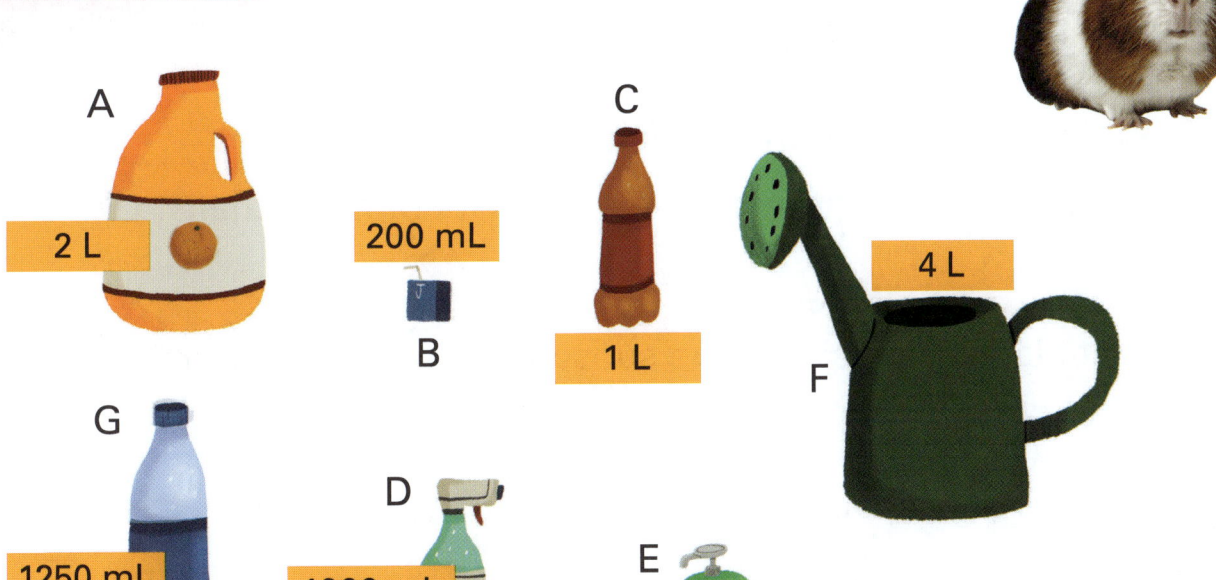

A 2 L

B 200 mL

C 1 L

F 4 L

G 1250 mL

D 1000 mL

E 500 mL

a   Write the letters of the items that hold less than 1 L.

_____

b   Write the letters of the items that hold more than 1 L.

_____

c   Write the letters of the items that hold exactly 1 L. _____

d   Which item has the biggest capacity? _____

e   Which item has the smallest capacity? _____

1

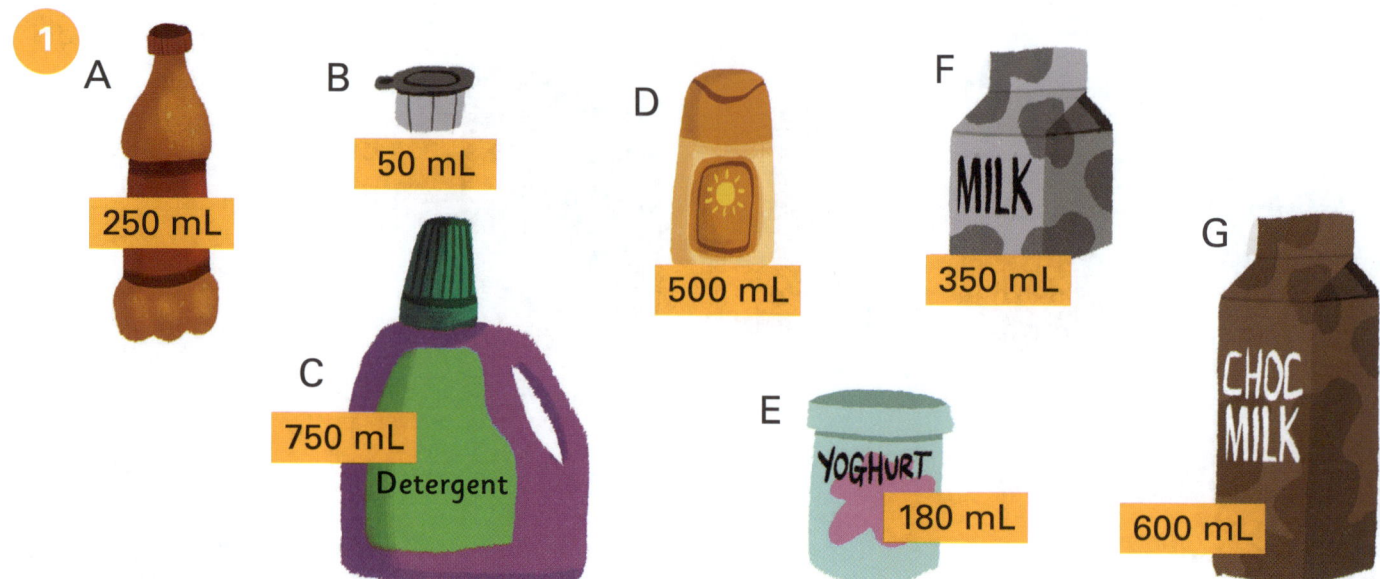

A 250 mL

B 50 mL

C 750 mL Detergent

D 500 mL

E YOGHURT 180 mL

F MILK 350 mL

G CHOC MILK 600 mL

a Which 2 items together have a capacity of 1 L? _____

b Which 2 items together have a capacity of more than 1 L?

_____

c What is the capacity of the sunscreen and the yoghurt?

_____

d What is the capacity of the detergent and the milk?

_____

2 You will need a 1 litre container.

a Choose 3 other containers and record them in the table below.

b For each container, estimate the capacity as more or less than 1 L.

c Use your 1 litre container to check. Record the results.

Container	I think it will hold …	It actually holds …
	more than 1 litre. less than 1 litre.	more than 1 litre. less than 1 litre.
	more than 1 litre. less than 1 litre.	more than 1 litre. less than 1 litre.
	more than 1 litre. less than 1 litre.	more than 1 litre. less than 1 litre.

OXFORD UNIVERSITY PRESS

Extended practice

1

a Use centicubes to make an object with a volume of 12 cm³.

b Draw your object.

2

a Use centicubes to make an object with a volume of 10 cm³.

b Draw your object.

3 Find 3 containers and record them in the table below.

a Estimate and record the capacity of each in mL.

b Measure and record the actual capacity in mL.

Container	Estimated capacity	Actual capacity

c Which container has the biggest capacity? _____

d Which has the smallest capacity? _____

The mass of lighter objects is measured in grams (g).

 15 grams or 15 g

There are 1000 g in 1 kg.

The mass of heavier objects is measured in kilograms (kg).

 15 kilograms or 15 kg

Is your mass closer to that of the cookie or the dog?

Guided practice

1 Write the item letters in order from lightest to heaviest.

a

 110 g **A**    125 g **B**   11 g **C**    410 g **D**    40 g **E**    250 g **F**

lightest ☐ ☐ ☐ ☐ ☐ ☐ heaviest

b

 2 kg **A**    4 kg **B**    32 kg **C**   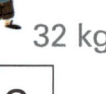 115 kg **D**    1 kg **E**    45 kg **F**

lightest ☐ ☐ ☐ ☐ ☐ ☐ heaviest

2

a Which item from question 1 is the heaviest? _____

b Which item is the lightest? _____

OXFORD UNIVERSITY PRESS

Independent practice

1 You will need a 1 kg weight.

a Choose 4 items in the classroom that you can easily pick up. Record them in the table below and tick whether you estimate each is heavier or lighter than 1 kg.

b Hold your 1 kg weight in one hand and heft each item in the other hand. Tick whether each item feels heavier or lighter than 1 kg.

Item	I think it is …		When I heft it feels …	
	lighter than 1 kg	heavier than 1 kg	lighter than 1 kg	heavier than 1 kg

c Check your 4 items using a pan balance and rewrite them in the correct columns below.

Lighter than 1 kg	Heavier than 1 kg

d Find and list 2 items that have a mass of about 1 kg.

_____   _____

2 You will need a 500 g weight.

a Choose and record 2 items that you think will have a mass of less than 500 g.

b Use a pan balance to check if they are less than or more than 500 g.

Item	Result	
	Less than 500 g	More than 500 g

What is the total mass of two 500 g weights?

c List 2 items that you think have a mass of about 500 g.

_____   _____

d Use a pan balance to check if the mass of your items is close to 500 g.

Circle the items that have a mass of around 500 g.

3 Find counters, blocks or other small objects.

Estimate and then check with a pan balance how many of your objects are needed to balance:

a a 10 g weight.   Estimate: _____   Actual: _____

b a 20 g weight.   Estimate: _____   Actual: _____

c a 50 g weight.   Estimate: _____   Actual: _____

4 How many 10 g weights do you need to balance:

a 20 g? _____   b 50 g? _____

c 100 g? _____   d 200 g? _____

e 150 g? _____   f 250 g? _____

OXFORD UNIVERSITY PRESS

1 Read the scales and record the mass.

a

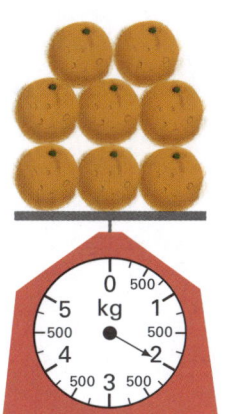

b

Mass: _____

Mass: _____

c

d

Mass: _____

Mass: _____

2 Look at the scales in question 1. What is the mass of:

a 1 orange? _____ g

b 1 pineapple? _____ kg

c 1 strawberry? _____ g

d 1 banana? _____ g

3 How much heavier are:

a the pineapples than the oranges? _____

b the bananas than the strawberries? _____

c the oranges than the bananas? _____

d the pineapples than the bananas? _____

The marks between each number on a clock represent 1 minute.

24 mins

The minute hand is pointing to the 36th minute so the time is **3:36** or **24 minutes to 4**.

36 mins

There are 60 minutes in 1 hour. Each of the numbers on the clock is 5 minutes apart. You can count by 5s to tell the time more quickly.

Guided practice

1 Write the analogue and digital times.

a

_____ past _____

[:]

b

_____ to _____

[:]

c

_____ past _____

[:]

d

_____

[:]

e

_____

[:]

f

_____

[:]

OXFORD UNIVERSITY PRESS

1 Draw in minute hands to show the times below.

a

$\frac{1}{4}$ to 5

b

7:23

c

5 past 5

d

12:14

e

14 to 12

f

11:59

2 Draw in hour hands to show the times below.

a

2:22

b

25 to 9

c

4:38

3 Draw in the hour and minute hands to show the times below.

a

11:11

b

9 past 7

c

7 past 9

d

8:44

e

11 to 2

f

2 to 11

4 How long did it take the minute hand to move:

a
from   to   **b**   from   to

_____   _____

c
from   to   **d**   from   to

_____   _____

> What takes about a minute to do? What takes 3 minutes?

5 How many minutes in:

a 1 hour? _____   **b** 2 hours? _____

c $\frac{1}{2}$ hour? _____   **d** $1\frac{1}{2}$ hours? _____

e $\frac{1}{4}$ hour? _____   **f** $\frac{3}{4}$ hours? _____

6 How many seconds in:

a 1 minute? _____   **b** 2 minutes? _____

c 5 minutes? _____   **d** 10 minutes? _____

e $3\frac{1}{2}$ minutes? _____   **f** $10\frac{1}{2}$ minutes? _____

OXFORD UNIVERSITY PRESS

1 Akira started brushing her teeth at 7:54. It took her 3 minutes.

a Mark the finish time on the clock.

b Write it in digital time.

c Write it in analogue time.

_____

2 Cian took 35 minutes to do his homework. He started at 4:47.

a Mark the start and finish times.

Start   Finish

b Write the finish time in digital time.

:

c Write it in analogue time.

_____

3 How many minutes until:

a 2:20? _____

b 2:48? _____

c 3:16? _____

d 3:00? _____

4 How long until:

a 8:00? _____

b 9:15? _____

c 7:55? _____

A shape is **regular** if all its sides are the same length.

Regular pentagon

Irregular shapes do not have all sides of equal length.

Irregular pentagon

The irregular pentagon has one pair of parallel sides and two right angles.

Guided practice

1   Match the quadrilaterals with their descriptions.

rectangle

parallelogram

rhombus

kite

trapezium

- regular shape
- type of parallelogram

- irregular
- 2 pairs of adjacent sides the same length

- irregular
- 1 pair of parallel sides

- irregular
- 2 pairs of parallel sides

- irregular
- 4 right angles
- 2 pairs of parallel sides

OXFORD UNIVERSITY PRESS

1 Complete the descriptions and name each shape.

a

Parallel lines: | Yes | No

Regular: | Yes | No

No. of sides: _____

Name: _____

b

Parallel lines: | Yes | No

Regular: | Yes | No

No. of sides: _____

Name: _____

c

Parallel lines: | Yes | No

Regular: | Yes | No

No. of sides: _____

Name: _____

d

Parallel lines: | Yes | No

Regular: | Yes | No

No. of sides: _____

Name: _____

e

Parallel lines: | Yes | No

Regular: | Yes | No

No. of sides: _____

Name: _____

2 Write 3 points to describe each shape, and then name it.

a

_____

_____

Name: _____

_____

b

_____

_____

Name: _____

_____

c

_____

_____

Name: _____

d

_____

_____

Name: _____

e

_____

_____

Name: _____

You can also think about corners and angles to help describe shapes.

OXFORD UNIVERSITY PRESS

1 You can make new shapes by joining 2 shapes together.

Draw lines to show the 2 shapes that join to make the shapes below. Then name them.

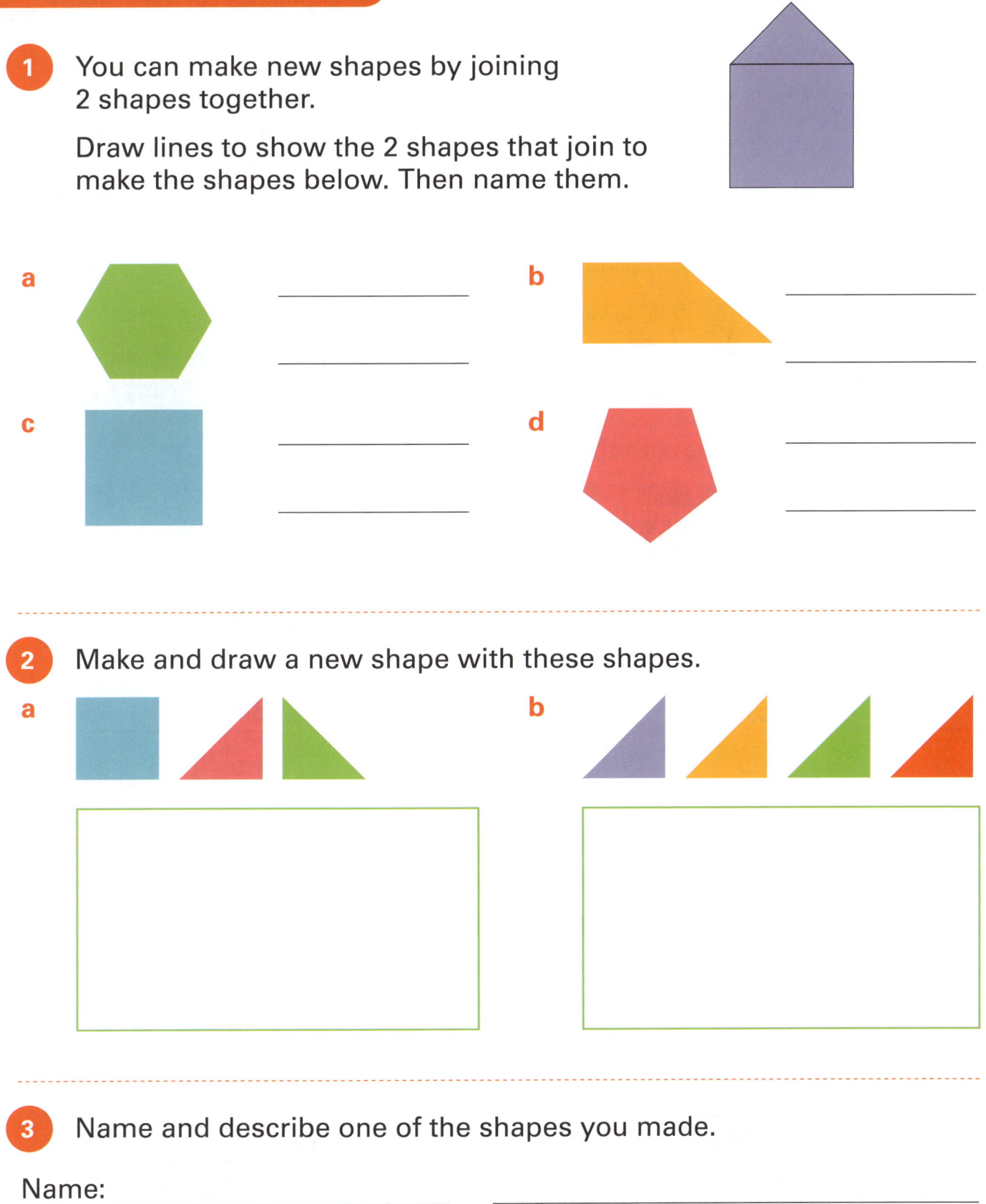

a _____

_____

b _____

_____

c _____

_____

d _____

_____

2 Make and draw a new shape with these shapes.

a

b

3 Name and describe one of the shapes you made.

Name: _____

_____

_____

_____

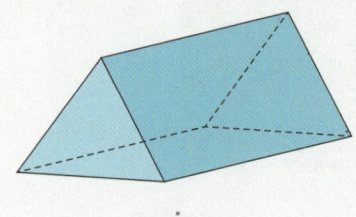 prism
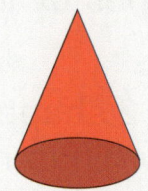 pyramid
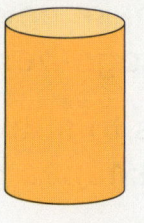 cone
 cylinder
sphere

> You can describe 3D shapes by their faces, edges and corners or vertices.

Guided practice

1 Match the objects with their descriptions.

cylinder   prism   sphere   pyramid   cone

• polygon as a base   • all other faces are triangles
• perfectly round 3D shape
• object with a circular base and a point at the tip
• 2 parallel bases the same shape   • all other faces are rectangles
• object with 2 circular ends and 1 curved face

2 Circle all the pyramids.

OXFORD UNIVERSITY PRESS

1

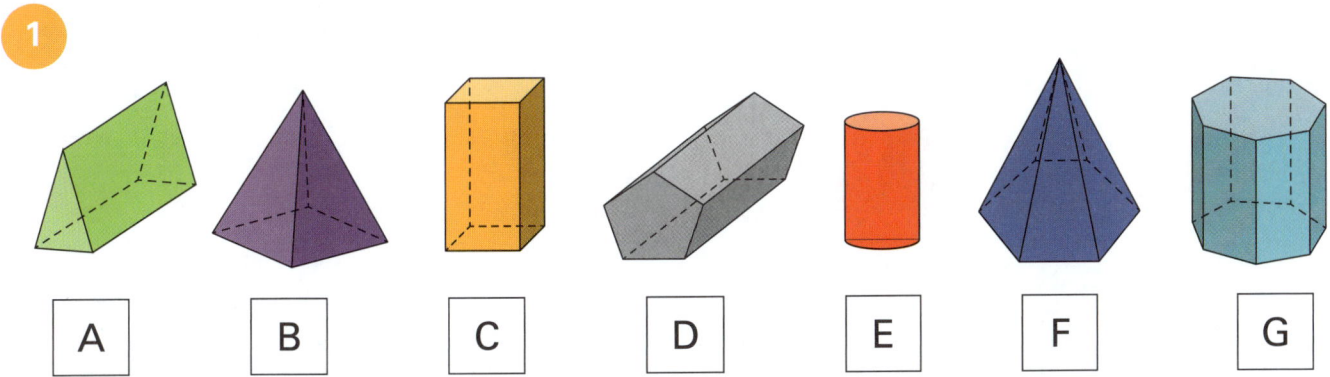

| A | B | C | D | E | F | G |

a Write the letters of the 3D shapes that are prisms.

_____

b Match the letters from question 1a to the descriptions of the pyramids below.

I have 10 corners and 15 edges. The shape of my bases has 5 sides.

All my faces are the same shape, but not the same size.
I have 8 corners.

I have 16 corners.
I have 10 faces.
I have 24 edges.

I have 5 faces.
I have 6 corners.
I have 9 edges.

c Draw a square prism.

d What is another name for a square prism?

_____

2 Circle all the 2D shapes you need to make the 3D shapes.

Make sure you circle 1 2D shape for every face of the 3D shapes.

a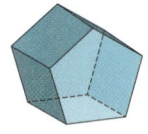

b

c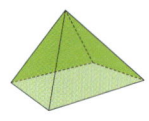

d

3 Write 1 similarity and 1 difference between these shapes.

a

Similarity: _____

_____

Difference: _____

_____

b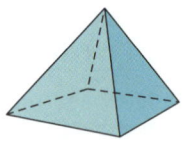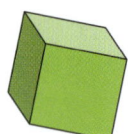

Similarity: _____

_____

Difference: _____

_____

c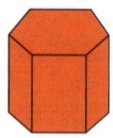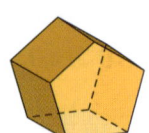

Similarity: _____

_____

Difference: _____

_____

d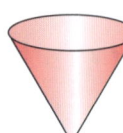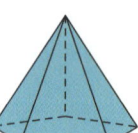

Similarity: _____

_____

Difference: _____

_____

OXFORD UNIVERSITY PRESS

Extended practice

When an object such as a box is flattened out, the 2D shape is called a **net**.

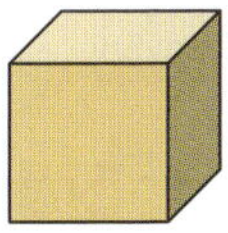

cube

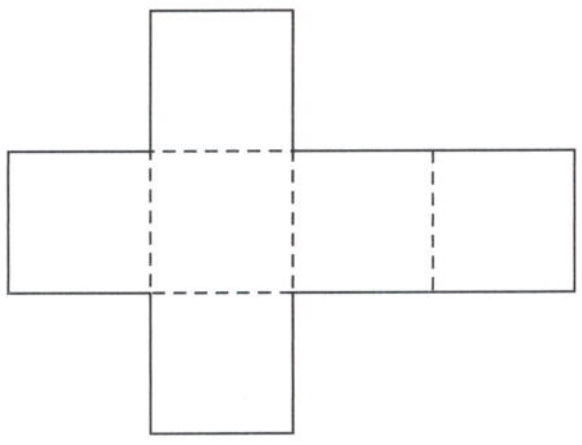

This is the net of a cube.

1 Match the nets to the 3D shapes.

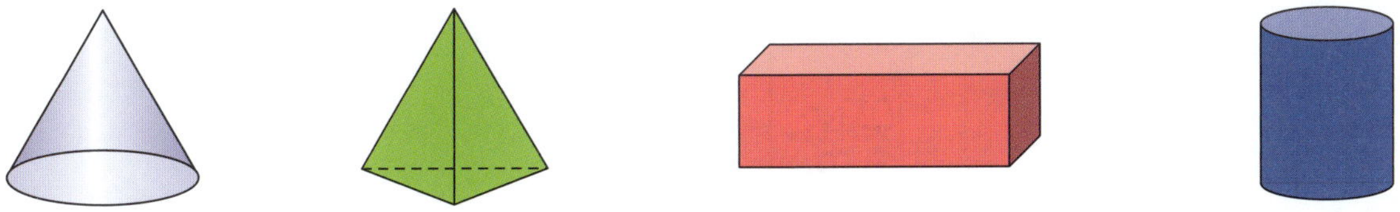

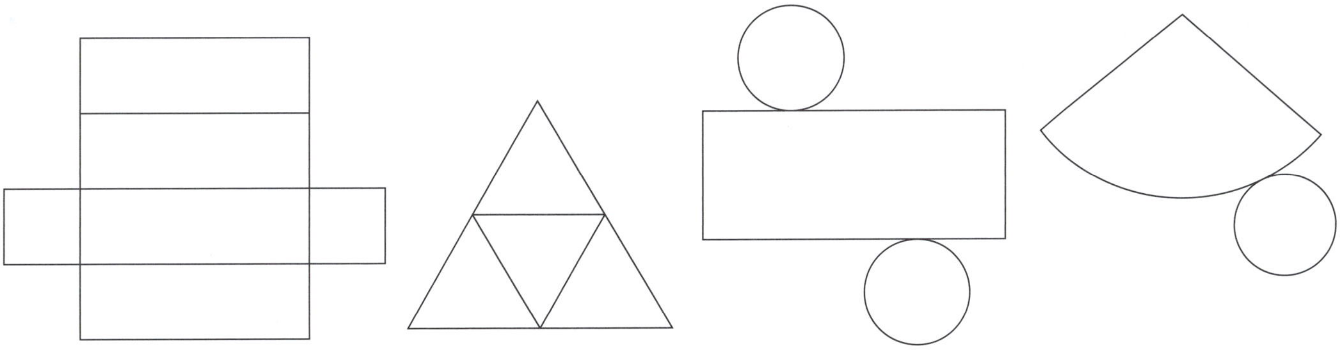

2 **a** Draw a prism.

b Name your prism.

c Write a description of your prism.

____

____

____

____

Name: ______________________________

An angle is the amount of turn between 2 arms.

A square corner angle is known as a right angle.

This angle is smaller than a right angle.

This angle is larger than a right angle.

The lines that make up an angle are called arms. The point where the 2 arms meet is the vertex.

Guided practice

1 Tick whether each angle is smaller or larger than a right angle.

a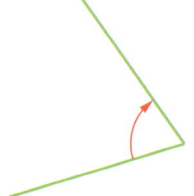
☐ smaller
☐ larger

b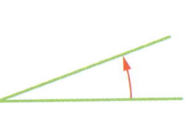
☐ smaller
☐ larger

c
☐ smaller
☐ larger

d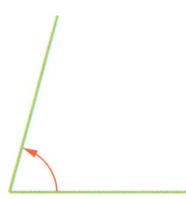
☐ smaller
☐ larger

e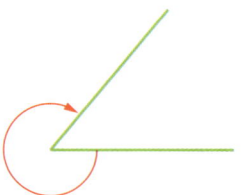
☐ smaller
☐ larger

f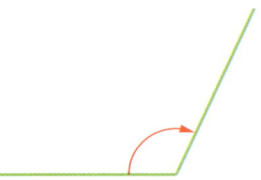
☐ smaller
☐ larger

OXFORD UNIVERSITY PRESS

1 Find and draw 3 things in your classroom that have a right angle.

---

2 Circle the shapes that have right angles.

a    b    c

d    e    f

---

3 How many right angles?

a    b    c

☐ right angles   ☐ right angles   ☐ right angles

4 Look at the angles marked between the clock hands.

| A | B | C | D | E | F |

a At what times do the hands make a right angle?

_____   _____

b Which clocks show angles smaller than a right angle?

_____

c Which clocks show angles larger than a right angle?

_____

What would the angle look like if it were 6 o'clock?

- -

5 **a** Draw your own times on each clock below.

b Draw a clockwise arrow to show the angle.

c Tick a box to classify each angle.

☐ Smaller than a right angle

☐ A right angle

☐ Larger than a right angle

☐ Smaller than a right angle

☐ A right angle

☐ Larger than a right angle

☐ Smaller than a right angle

☐ A right angle

☐ Larger than a right angle

OXFORD UNIVERSITY PRESS

Extended practice

1 **a** Find and draw 4 angles in the classroom.

  b Write a description to classify your angle compared to a right angle.

Angle 1

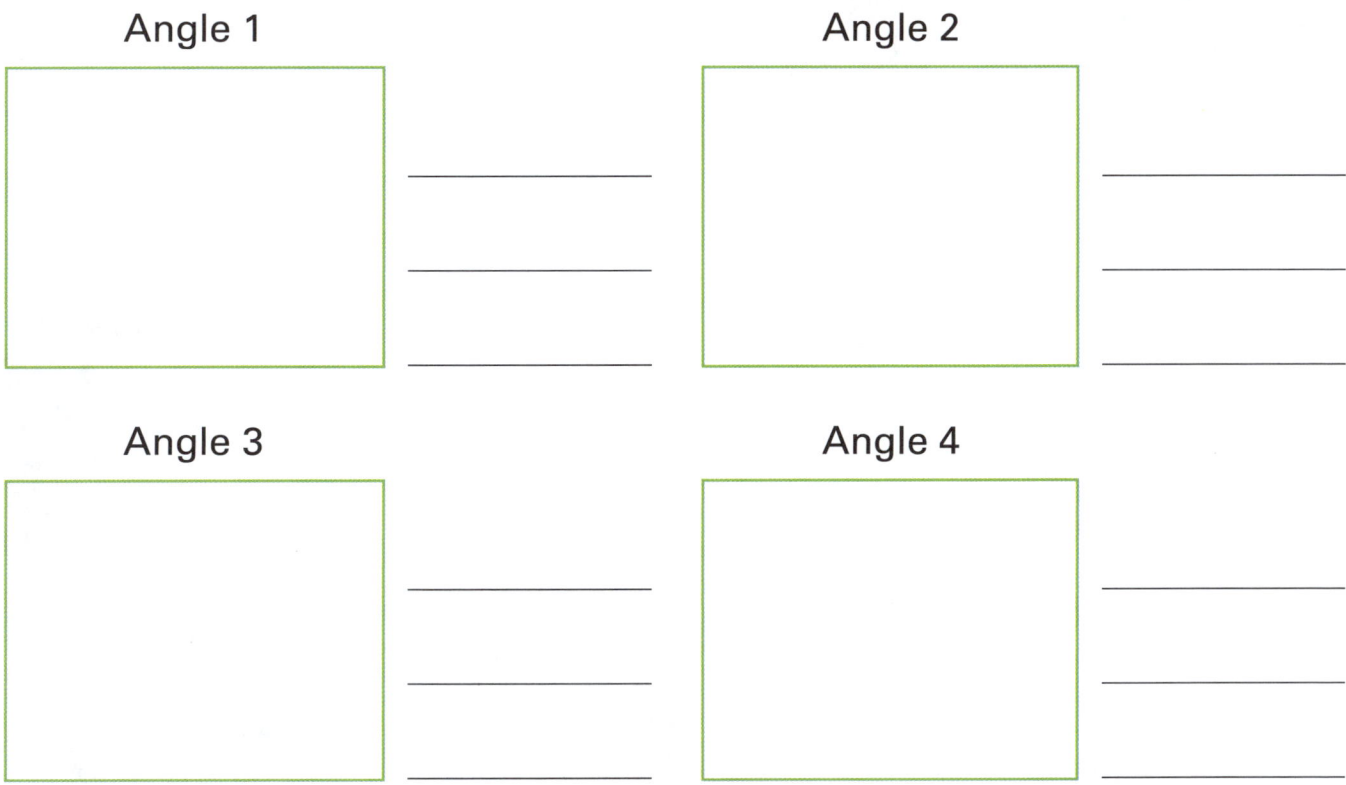

_____

_____

_____

Angle 2

_____

_____

_____

Angle 3

_____

_____

_____

Angle 4

_____

_____

_____

2 Draw lines to match the angles that are the same size.

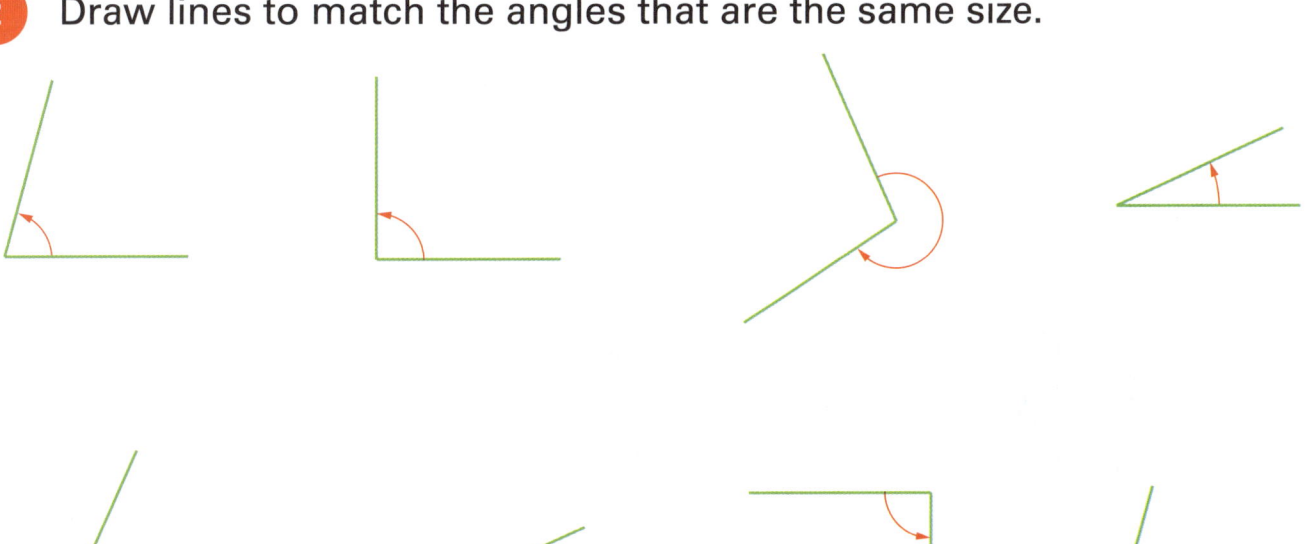

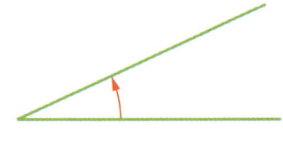

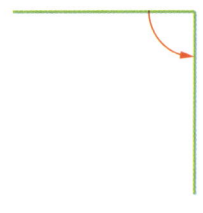

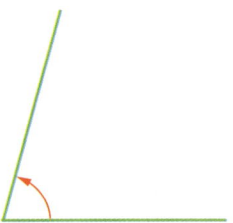

An object is symmetrical if one side is a mirror image of the other.

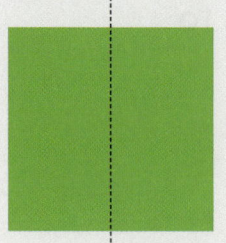

This square is symmetrical.

This butterfly is symmetrical.

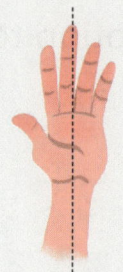

This hand is not symmetrical.

Line symmetry can be horizontal, vertical or even diagonal.

Guided practice

1 Tick if each item is symmetrical or not.

a

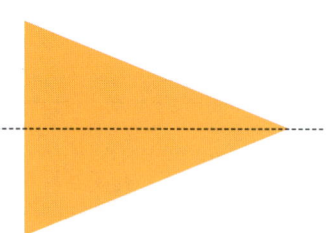

☐ Symmetrical

☐ Not symmetrical

b

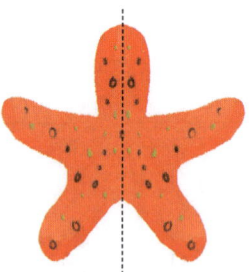

☐ Symmetrical

☐ Not symmetrical

c

☐ Symmetrical

☐ Not symmetrical

d

☐ Symmetrical

☐ Not symmetrical

e

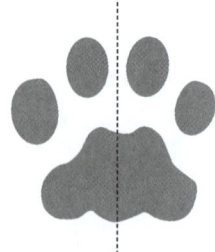

☐ Symmetrical

☐ Not symmetrical

f

☐ Symmetrical

☐ Not symmetrical

OXFORD UNIVERSITY PRESS

1 Draw 1 line of symmetry on each shape.

a  b  c

d  e  f

2 Draw 2 lines of symmetry on each shape.

a  b  c

d 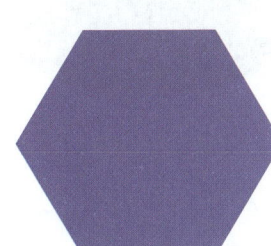 e  f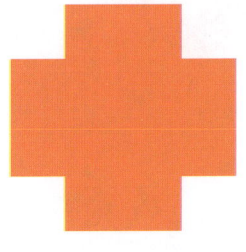

3 a Which shape in question 2 has exactly 3 lines of symmetry?

_____

b Which shapes have exactly 4 lines of symmetry?

_____

4 **a** Find and draw 4 symmetrical items.

    b Draw a line of symmetry on each.

---

5 Circle the shapes with line symmetry.

a

b

c

d

e

Are you symmetrical?

OXFORD UNIVERSITY PRESS

1   Draw a symmetrical shape picture.

2   Create a symmetrical picture on the grid.
Make sure that one side of the picture
is a reflection of the other side.

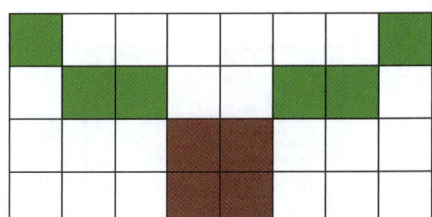

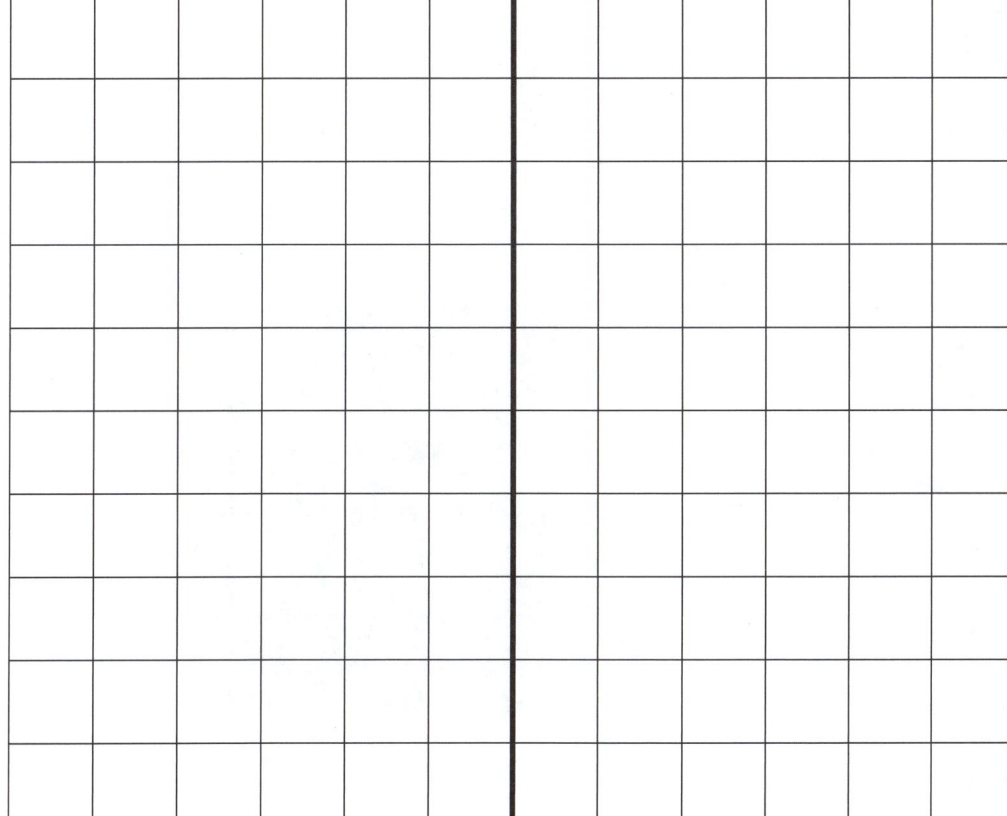

There are examples of slides and turns all around us.

This brick pattern shows slides.

This brick pattern shows turns.

What different meanings does the word "slide" have?

Guided practice

1   Slide or turn?

a

| Slide | Turn |

b

| Slide | Turn |

c

| Slide | Turn |

d

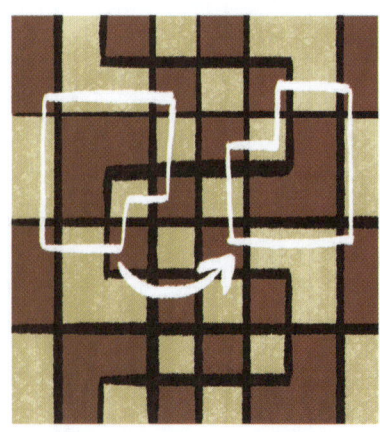

| Slide | Turn |

OXFORD UNIVERSITY PRESS

1 Follow the rules to make repeating patterns.

a slide, then quarter turn clockwise

b half turn, then quarter turn clockwise

c half turn, then slide

d quarter turn, then half turn anticlockwise

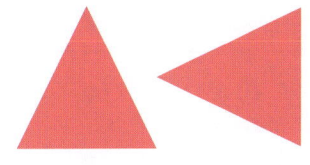

e half turn, slide, then quarter turn anticlockwise

2　**a**　Make your own slide and turn pattern.

b　Write the rule for your pattern.

_____

_____

A flip is when an object is turned over to be a mirror image of itself.

3　Slide, turn or flip?

a

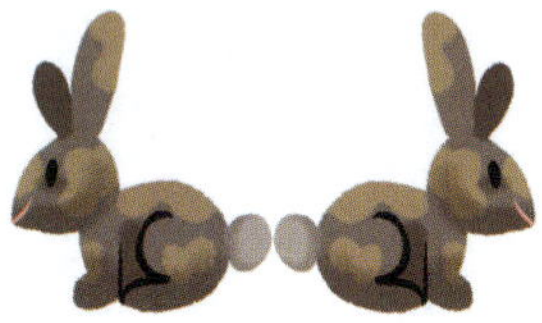

_____

b

_____

c

_____

4　Find 2 examples of flip, slide or turn patterns in your classroom.

a　Draw each pattern.

b　Label the translations.

OXFORD UNIVERSITY PRESS

1 Circle and label slides, turns and flips in these designs.

a

b

c

d

2 Design your own T-shirt patterns using slides, turns and flips.

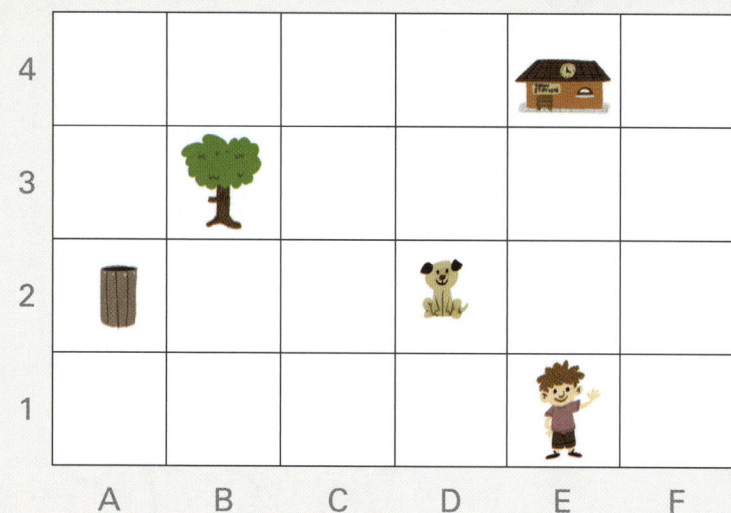

The tree is at B3.

The boy is at E1.

The station is at E4.

To find what is at D2, put one finger on D and another on 2 and move them along the lines until they meet.

Guided practice

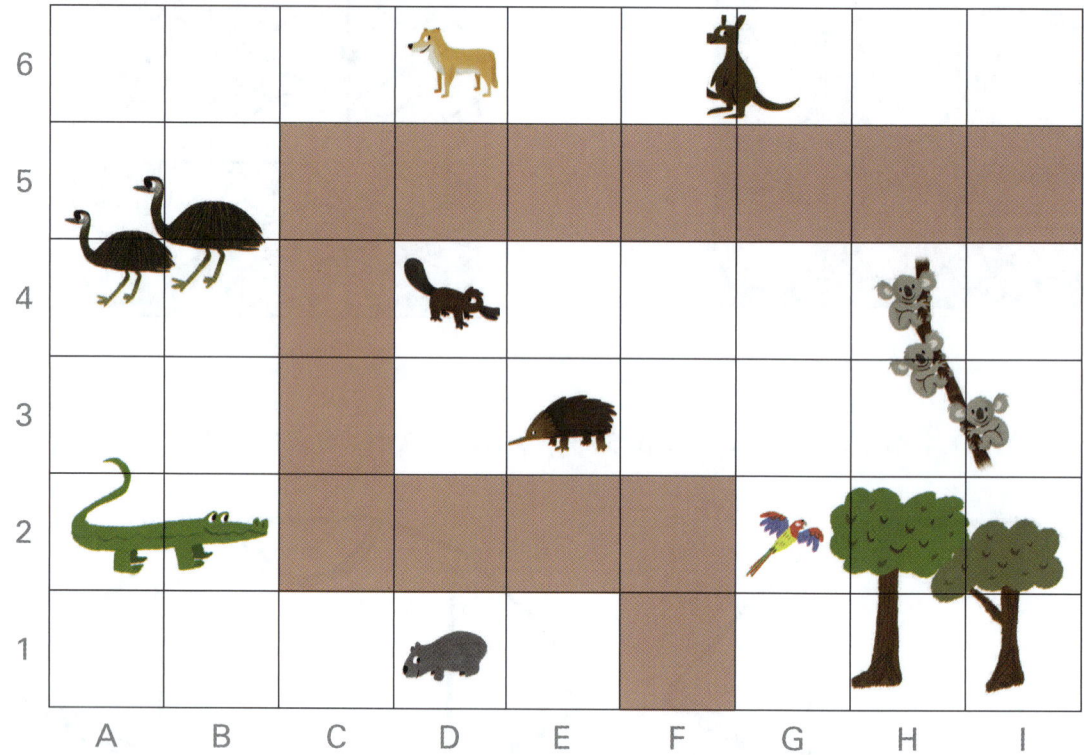

1 What is at:

a D1? _____

b D6? _____

c G2? _____

d A2? _____

e D4? _____

f H4? _____

OXFORD UNIVERSITY PRESS

1 Write the letters in the correct squares.

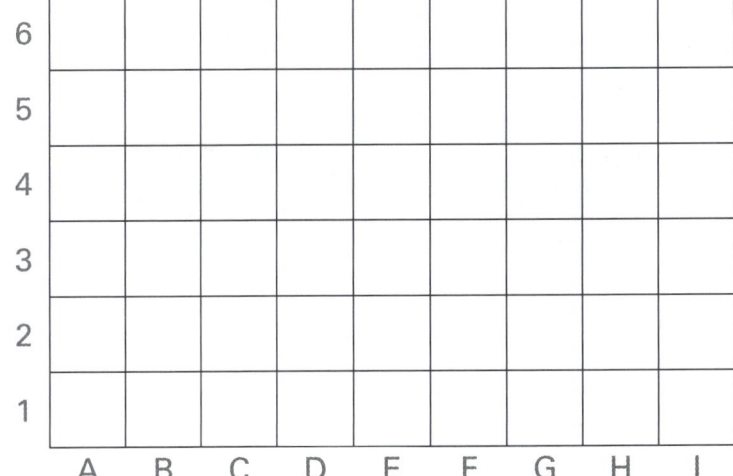

a   E in C4

b   L in E2

c   N in H4

d   L in D3

e   D in F4

f   W in B5

g   O in G3

h   E in I5

2 Write the grid references for these locations.

a   the entrance:

_____

b   the hotdog stand:

_____

c   the dodgem cars: _____ and _____

d   the roller coaster: _____ , _____ , _____ and _____

e   the pirate swing: _____ and _____

f   the Ferris wheel: _____ and _____

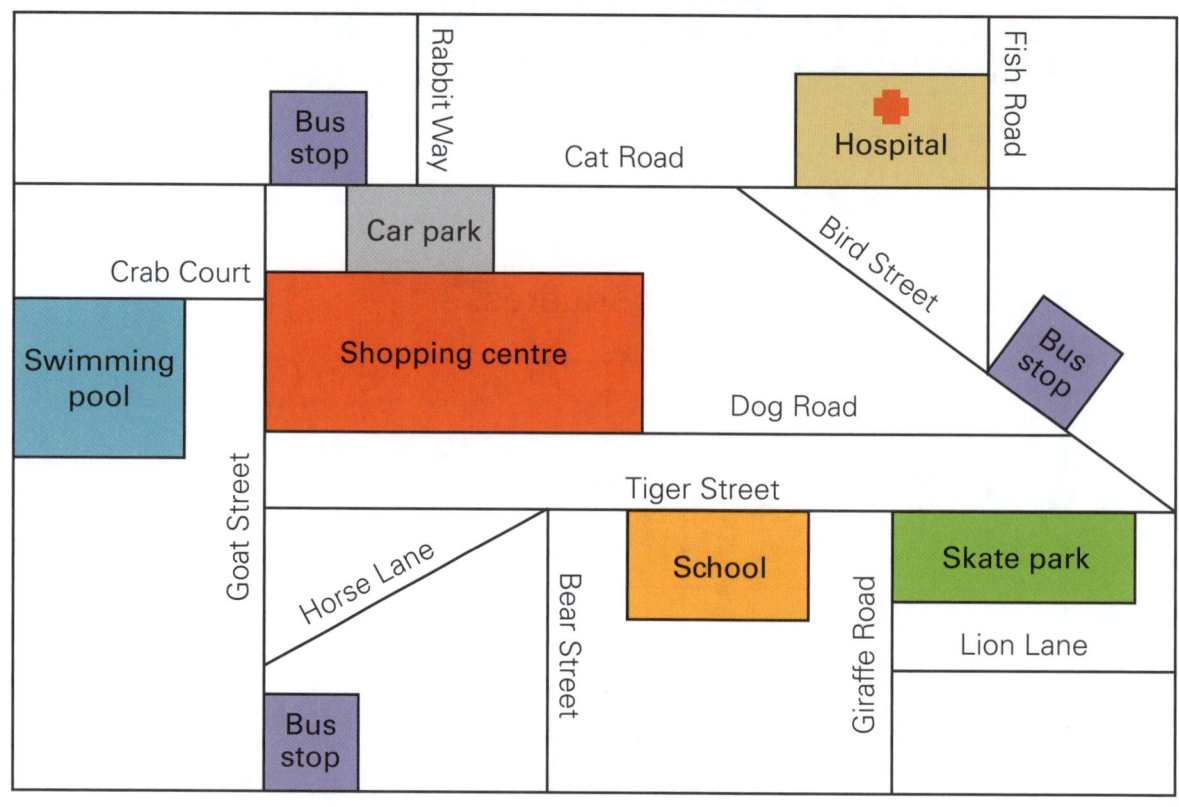

3 **a** Which 2 roads is the skate park on?

_____

b Which 2 roads is the hospital on?

_____

4 Follow the directions.

a Start at the Bird St bus stop.

b Walk along Bird St to Cat Rd.

c Turn left onto Cat Rd.

d Keep walking until you reach Goat St.

e Turn left and walk to the corner of Dog Rd.

f Where are you now? _____

Remember to consider where you are on the map when turning left or right.

5 Write your own directions from the swimming pool to the school.

_____

_____

_____

OXFORD UNIVERSITY PRESS

1 Create a map of your classroom or school.

2 Write directions from one place on your map to another.

_____

_____

_____

3 This is a bird's-eye view of a park.

5

4

3

2

1

A   B   C   D   E   F

Write the grid reference for:

a   a tree. _____

b   the picnic table. _____

c   the slide. _____

d   the ducks. _____

You can collect data from many different sources.

observation

surveys

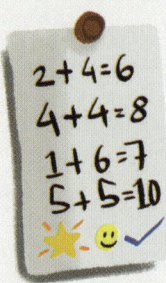
test results

other sources

Which source of data might be best for finding out where your class likes to go for holidays?

Guided practice

1 Match the data with the best source.

favourite food in your class	favourite food in your country	number of people who walked past the school during lunch	students in your class who know their times tables
observation	survey	test results	other sources, such as government department of statistics

2 What answers might you expect if you asked your classmates about:

a their favourite sport? _____

b their favourite colour? _____

c what pets they have? _____

OXFORD UNIVERSITY PRESS

1

a Write a survey question to find out about your classmates' hobbies.

_____

b Ask 10 people your question and record their responses in the table.

Responses	Number of people									
	1	2	3	4	5	6	7	8	9	10

2 Circle the best question to ask if you want to find out the number of brothers and sisters your classmates have.

a Do you have any brothers and sisters?

b How many people in your family?

c How many brothers and sisters do you have?

3 Ask 5 people the question you chose and record their answers with ticks.

	0	1	2	3	4 or more
Number of brothers and sisters					

4 The data in this list was collected in a survey. Reorganise the data as a table using tally marks.

Survey question: What is your favourite colour?

List

blue, red, blue, green, red, red, green, blue, pink, red, blue, red

Table

Colour	Responses

5 Survey 12 people in your class about their favourite animal.

a Write the question you will ask them.

_____

b List their responses.

c Show their responses in a table.

OXFORD UNIVERSITY PRESS

1 These shapes have been sorted into 3 groups.

a Explain how they are grouped.

_____

b What source do you think was used to classify the data?

_____

2

a Choose and record one type of data you could collect in the classroom through observation.

_____

b Collect and record the data in a list or table.

This is a bar graph.

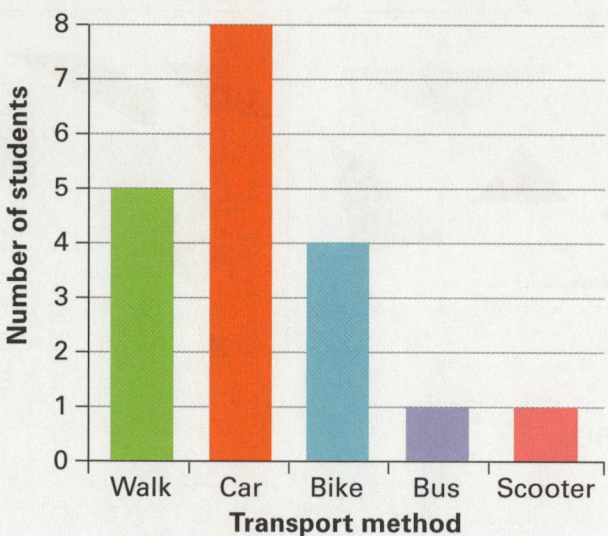

The x-axis is also called the horizontal axis, and the y-axis is called the vertical axis.

Guided practice

1

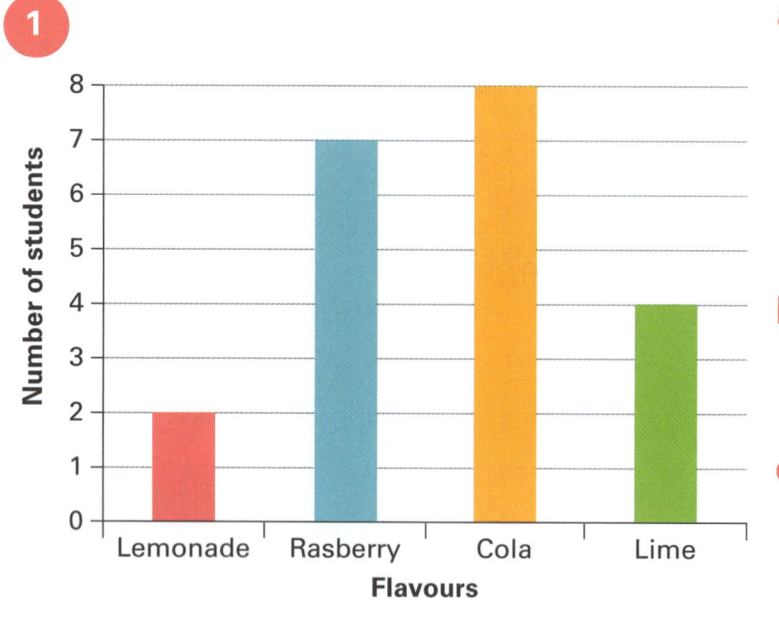

a What is the title of the graph?

_____

_____

b What does the *x*-axis show?

_____

c What does the *y*-axis show?

_____

d How many different flavours are recorded? _____

e What is the highest number on the *y*-axis? _____

f Which flavour was the favourite of the least number of students?

_____

OXFORD UNIVERSITY PRESS

1   This table shows the favourite day of the week in class 3S.

Day	Monday	Tuesday	Wednesday	Thursday	Friday	Saturday	Sunday
Number of students							
	I	III	II	II	IIII	ⅢⅢ I	Ⅲ II

a   Use the data to complete the graph.

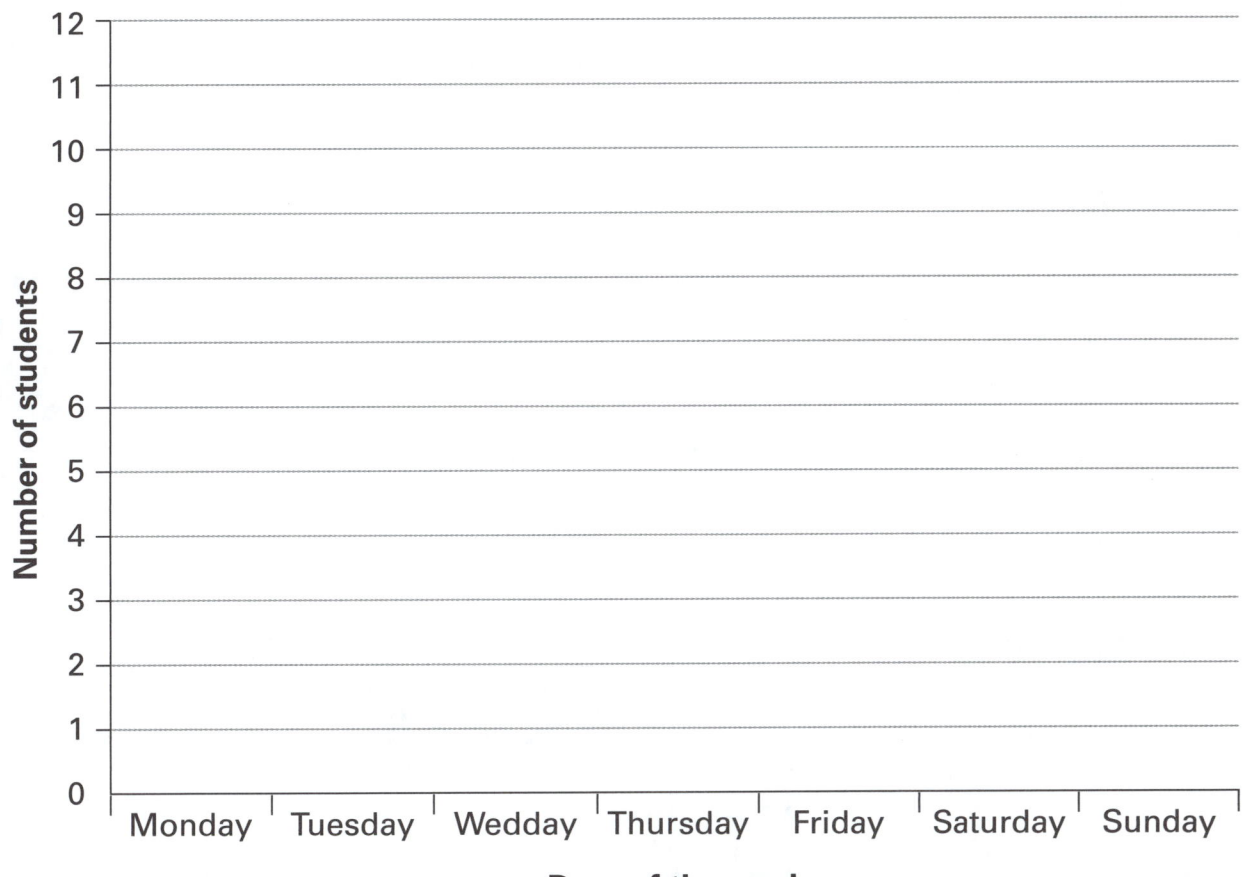

b   Which day is the most popular? _____

c   Which day is the least popular? _____

d   What does the *x*-axis show? _____

e   What does the *y*-axis show? _____

f   What was the highest total recorded? _____

2

a Survey 10 classmates about their favourite meal and record the data as a list.

b Make a pictograph with the data.

Number of people									
Breakfast									
Lunch									
Dinner									

How is a pictograph different from a bar graph? How are they the same?

c Which meal was the most popular? _____

d How many people preferred breakfast? _____

3 Make a table with tally marks using the bar graph data.

WHERE I WAS BORN

Number of people
(Italy: 1, New Zealand: 4, Australia: 8, Vietnam: 2)
Country

WHERE I WAS BORN

	Country		
Number of people			

OXFORD UNIVERSITY PRESS

1 Survey 15 classmates to find out their birth order in their family.

a Record your results in the table.

Position	1st	2nd	3rd	4th	5th	6th or more
Number of students						

b Make a pictograph with the results.

1st	
2nd	
3rd	
4th	
5th	
6th or more	

c Make a bar graph with the results.

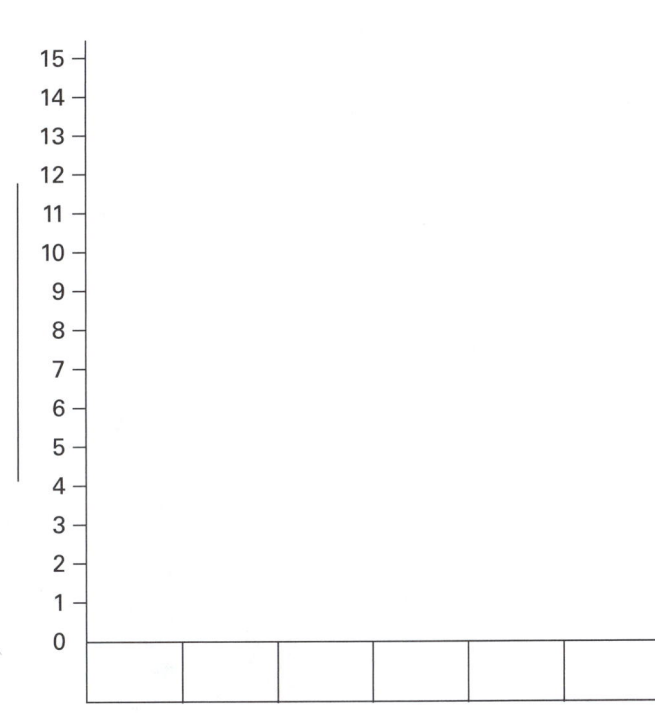

d Give both graphs a title and labels.

e Which graph do you find easier to read? Why?

_____

_____

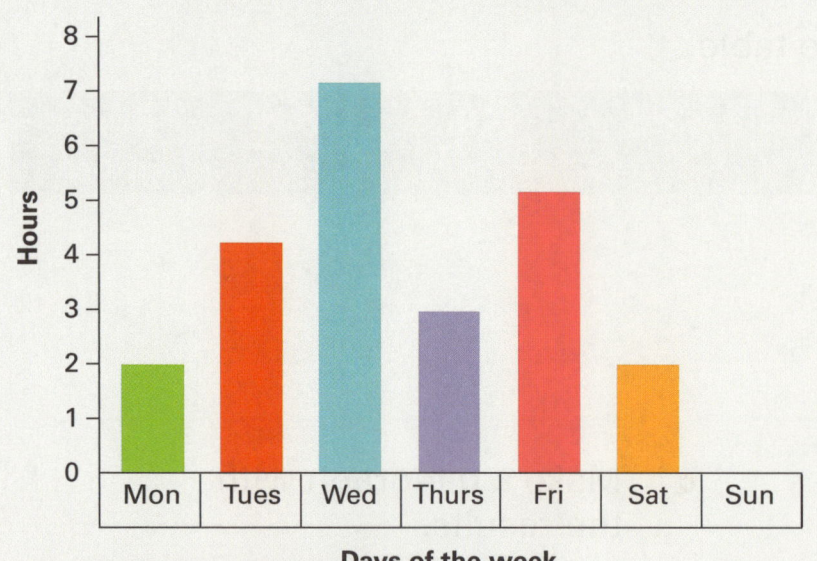

HOURS OLEG SPENT TRAINING THIS WEEK

- Oleg did the most training on Wednesday.
- He didn't do any training on Sunday.
- He did 2 hours of training on Monday.

What else does the graph tell you?

Guided practice

1 Use the data to answer the questions.

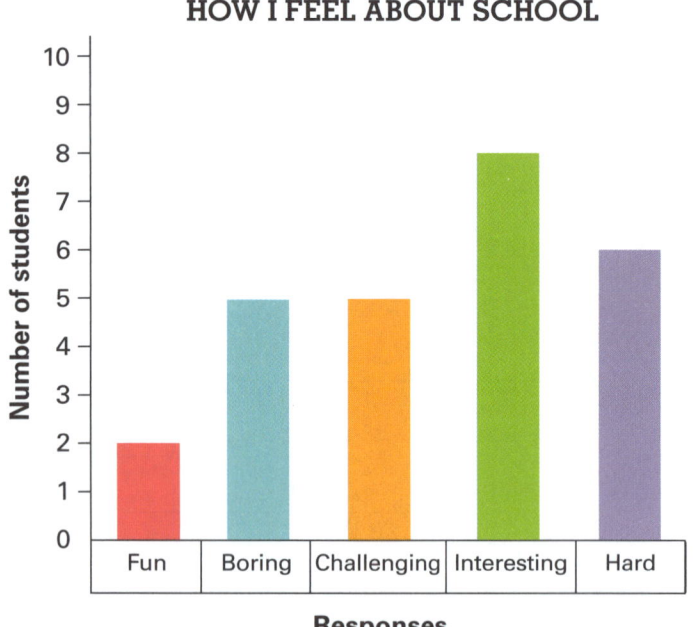

HOW I FEEL ABOUT SCHOOL

a Which response was most popular?

_____

b Least popular?

_____

c Which response did 6 students choose?

_____

d Which 2 responses were chosen by the same number of students?

_____   _____

e How many students were surveyed? _____

OXFORD UNIVERSITY PRESS

Independent practice

1 Use the data to answer the questions.

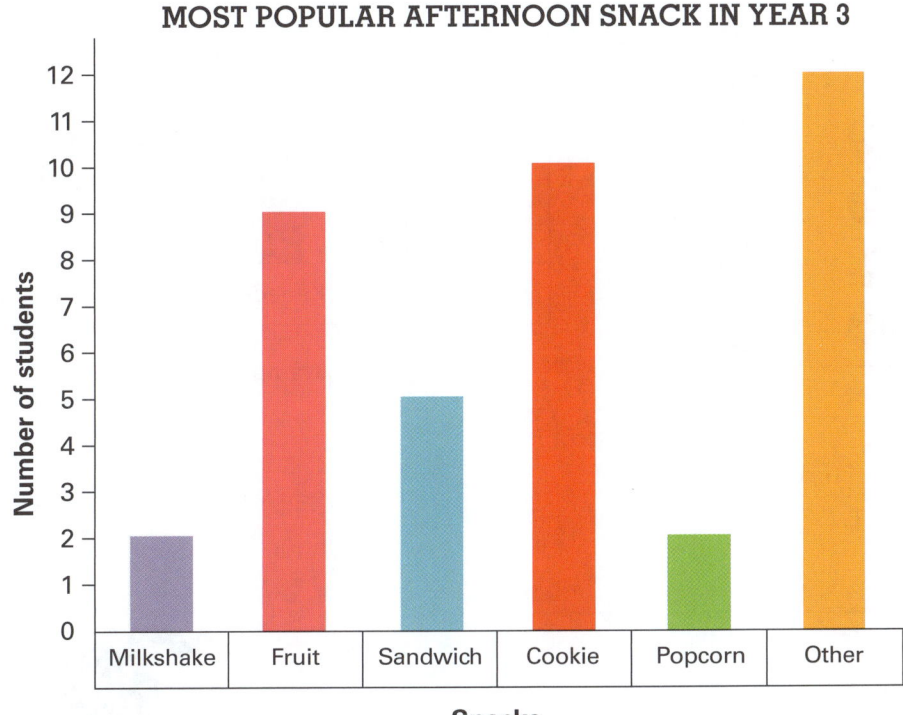

MOST POPULAR AFTERNOON SNACK IN YEAR 3

a How many more students chose fruit than popcorn? _____

b Did more students choose milkshakes or cookies? _____

c What might "Other" be? _____

2 Write 4 more statements about the data on the graph.

_____

_____

_____

_____

_____

3 These graphs show how many goals 5 students scored in a football season.

GOALS SCORED IN SEASON

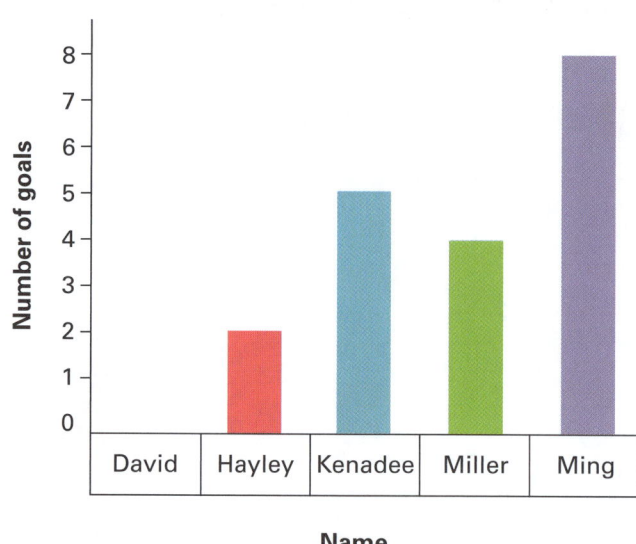

| David | Hayley | Kenadee | Miller | Ming |

a List 2 features the bar graph has that the pictograph doesn't.

_____

_____

_____

_____

How do you know which one is the bar graph?

b When might you use the first type of graph?

_____

c When might you use the second type of graph?

_____

d Write 2 facts from the data in the graphs.

_____

_____

e How many more did the highest goal scorer score than the lowest?

_____

f How many goals did the students score altogether in the season?

_____

OXFORD UNIVERSITY PRESS

1

a   Choose a survey topic (such as favourite foods) and write a question to ask your classmates.

Topic: _____   Question: _____

b   Survey 12 students and record their responses.

c   Make a graph of the results.

d   Write 3 statements about your data.

_____

_____

_____

We use diagrams to sort information in different ways.

We could use a **Venn diagram**.

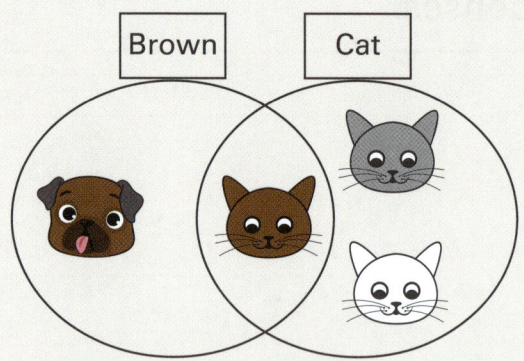

We could use a **Carroll diagram**.

How else could they be sorted in the diagrams?

Guided practice

1   **a**   Look at the Venn diagrams. Sort the cats and dogs from above into the correct places.

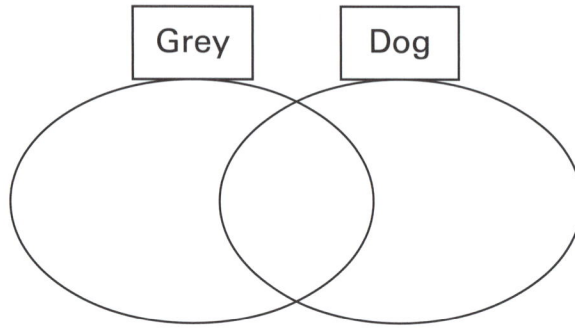 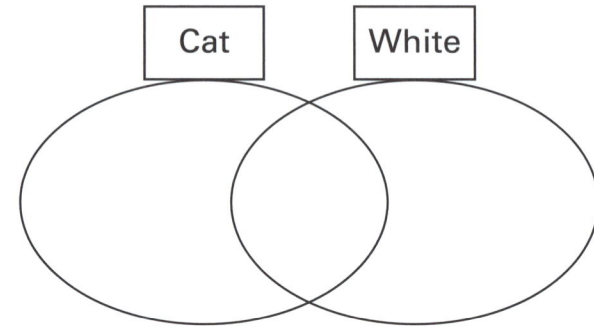

b   Sort the cats and dogs into the correct places in the Carroll diagrams.

	Cat	Dog
White		
Not white		

	Grey	Not grey
Cat		
Dog		

OXFORD UNIVERSITY PRESS

Independent practice

Look at these 2D shapes.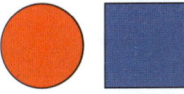

1

a Sort the shapes into groups in the Venn and Carroll diagrams.

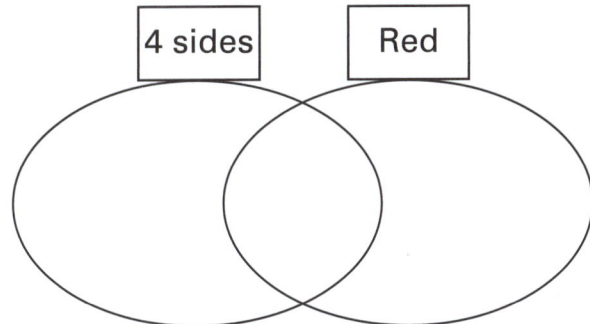

	4 sides	Not 4 sides
Blue		
Not blue		

b Which shapes are not blue and are not 4-sided? _____

c Which shape is red and 4-sided? _____

2 Look at how the shapes have been sorted. Write labels on the Venn and Carroll diagrams.

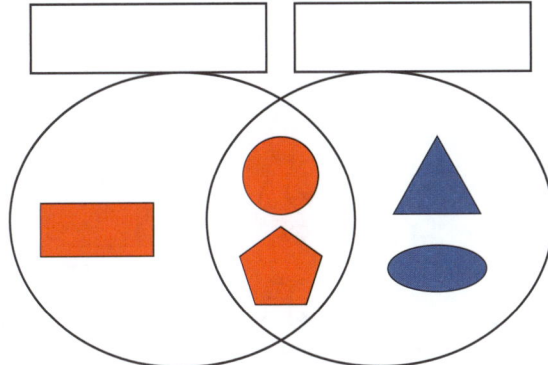

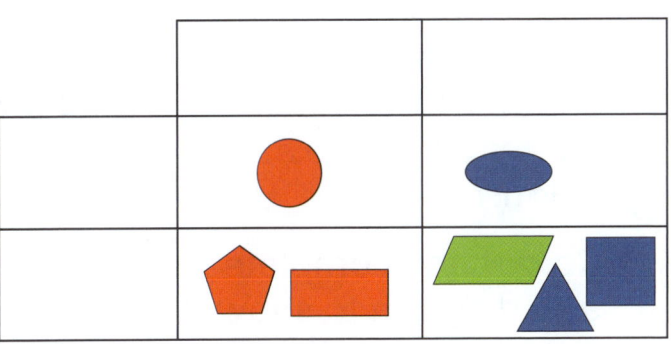

3 Complete the Venn diagram using the numbers.

30	18	23	11
15	12	13	5
21	6	27	3

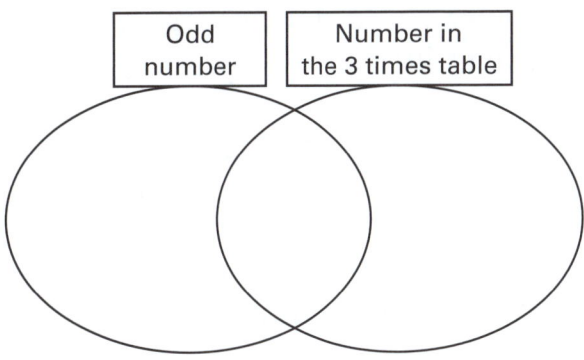

4 If you toss a coin and call "heads", you might be right and you might be wrong. Could you get "heads" twice in a row?

We can use a **tree diagram** to show the chance of this happening.

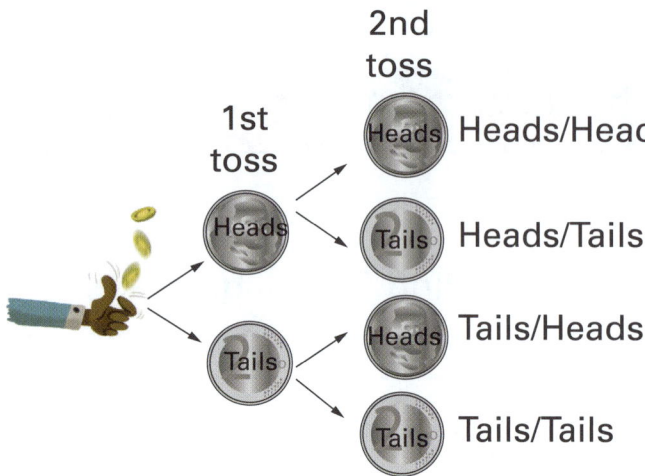

2nd toss

1st toss

Heads/Heads

Heads/Tails

Tails/Heads

Tails/Tails

There are four possible outcomes:

- heads then heads

- heads then tails

- tails then heads

- tails then tails.

There is one chance for heads and heads. That means that there is a quarter of a chance of getting heads twice.

a What fraction of a chance does tails and tails have? _____

b What fraction of a chance does heads and tails have? _____

---

5 This is Billy's sock game. In a box, there are four socks – two are red and two are blue.

Billy's mother blindfolds him and says, "Take out one sock and then another." Can he get a pair of socks the same colour?

Possible outcomes

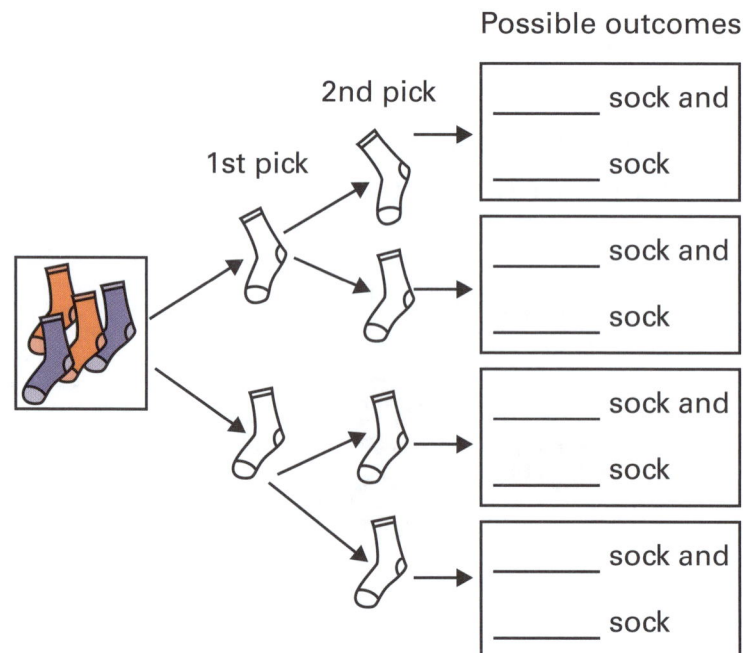

2nd pick

1st pick

_____ sock and _____ sock

_____ sock and _____ sock

_____ sock and _____ sock

_____ sock and _____ sock

a Colour and complete the tree diagram to show the possible outcomes

b Circle the correct answer below. The chance of getting a blue pair of socks is:

less than a red pair.   more than a red pair.   the same as a red pair.

c What fraction of a chance is there for a pair of red socks? _____

d Explain why there is more chance of getting an odd pair than a blue pair. _____

OXFORD UNIVERSITY PRESS

1

a Complete the diagrams using the numbers below from various multiplication tables.

50   24   6   35   36   10   21   40   18   12

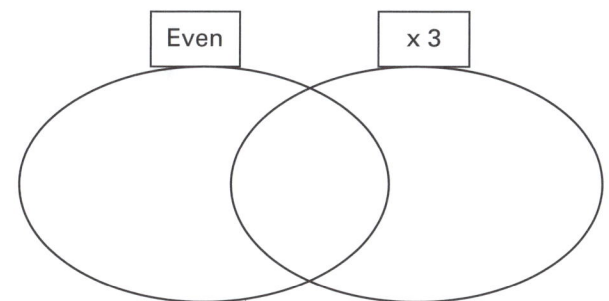

	x 5	x 3
x 2		
x 7		

b Write another number that could go in the place where the two circles overlap. _____

c What other number could go in the same space as 21? _____

2 Year 3 is having a special hat day. The students can choose a red, blue or yellow hat. They can decorate it with a flower, a star or a smiley face.

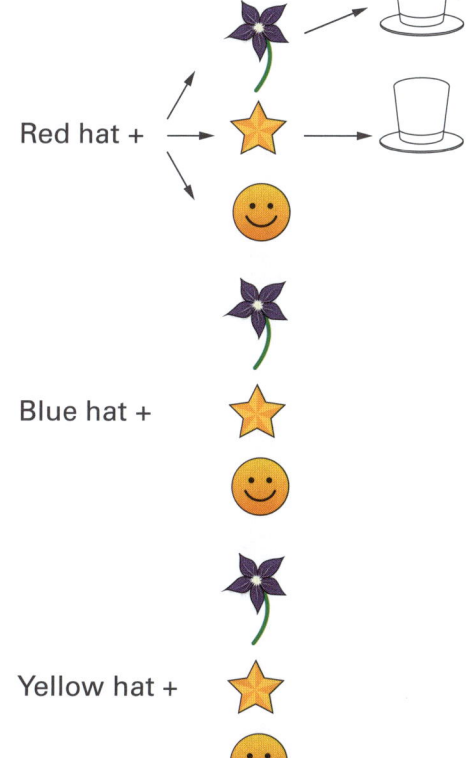

Red hat +

Blue hat +

Yellow hat +

a Complete the tree diagram to show the different hats that the students could make.

b How many different types of hats could there be? _____

c There are 36 children in Year 3. How many hats are likely to be red and have a flower? _____

If you have 2 ice-cream flavours and 2 toppings, these are the combinations you could make:

vanilla with sprinkles

vanilla with nuts

strawberry with sprinkles

strawberry with nuts

*The possible combinations can also be called **outcomes**.*

Guided practice

1

a Predict how many outcomes would be possible with 3 flavours and 2 toppings. _____

b Draw or write each of the combinations.

c How many are there? _____

OXFORD UNIVERSITY PRESS

1

a Jawad put a red, a blue, a green and a yellow marble in a box. List the possible outcomes if he draws out 2 of them at once.

b How many possible outcomes do you think there will be if he adds a purple marble? _____

c List or draw all the possibilities.

d How many are there? _____

e How likely is it that Jawad draws out a red marble on the first try?

impossible   less likely   most likely   certain

f How likely is it that he draws out a black one?

impossible   less likely   most likely   certain

2

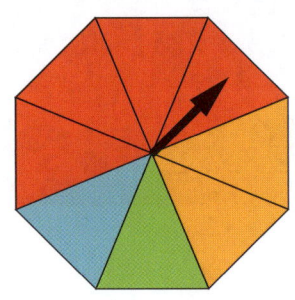

a How many different outcomes are possible on this spinner?

_____

b How likely is it to land on:

   i   red? _____   **ii**   green? _____

   iii pink? _____   **iv** yellow? _____

c What is the arrow most likely to land on?

_____

d What is the arrow least likely to land on?

_____

When might you need to know how likely something is?

3 Colour the spinner so that:

a it is most likely to land on green.

b it is least likely to land on blue.

c it is impossible to land on yellow.

d it is possible to land on red.

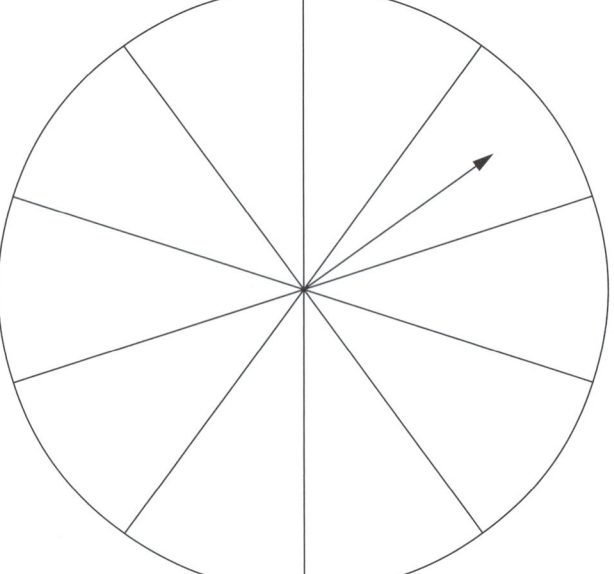

4 How many outcomes are possible if you toss:

a 1 coin? _____

b 2 coins? _____

c 3 coins? _____

5 Why do you think people use tossing coins to make decisions?

_____

OXFORD UNIVERSITY PRESS

1 Imagine a box containing 1 red and 1 blue counter. If you draw the counters out of the box one-by-one, 2 outcomes are possible:

red blue   **or**   blue red

a Predict how many combinations are possible if there are 3 colours.

_____

b Draw or list the possible outcomes if a pink counter is added.

2 Write 3 likelihood statements about the gumballs in this machine.

1. _____

_____

2. _____

_____

3. _____

_____

After 10 rolls of a dice, Penny recorded the following results.

Outcome	1	2	3	4	5	6
Number of times	||			||||	||			

If Penny rolls again, what do you think the next number will be?

Guided practice

1 Now it's your turn.

a Predict what your results will be if you roll a dice 10 times.

Outcome	1	2	3	4	5	6
Predicted number of times						

b Conduct the experiment and record the results.

Outcome	1	2	3	4	5	6
Actual number of times						

c Was your prediction correct? _____

d Why or why not?

_____

_____

OXFORD UNIVERSITY PRESS

1

a   Roll a dice 30 times and record the results.

Outcome	1	2	3	4	5	6
Number of times						

b   If you repeat the experiment, do you think the results will be the same? Why or why not?

_____

_____

c   Roll a dice another 30 times.

Outcome	1	2	3	4	5	6
Number of times						

d   Were the results different? Why or why not?

_____

_____

e   What would you expect if you did the experiment again?

_____

_____

f   How might the results be different if you repeated the experiment with a 10-sided dice?

_____

_____

2

a What are the 2 possible outcomes if you toss a coin?

_____   _____

b What are the 4 possible outcomes if you toss 2 coins?

_____   _____

_____   _____

c How likely are you to toss 2 heads rather than the other outcomes?

     less likely   equally likely   more likely

d Conduct 20 trials and record the results.

Outcome	Tail/tail	Tail/head	Head/tail	Head/head
Number of times				

e Which outcome came up most often?

_____

f Which came up least often? _____

g Do you think your results are the same as

other people in your class? _____

h Compare your results with a classmate.
What do they tell you about chance?

_____

_____

Have you ever made a decision by tossing a coin?

---

3 Circle the activities in which chance plays a part.

- winning a raffle
- catching a cold
- getting a perfect score on a spelling test
- going to the movies with your friends

OXFORD UNIVERSITY PRESS

1   Put 5 different coloured counters into a container.

a   If you take out 1 counter, what colour do you think it will be? Why?

_____

_____

b   Conduct the experiment 25 times, returning the counters to the box each time. Complete the table and record your results.

Outcome					
Number of times					

c   Make a pictograph of the results.

COUNTER EXPERIMENT OUTCOMES

Number of people

Colour

d   Write 2 statements about the results.

1.   _____

_____

2.   _____

_____

GLOSSARY

acute angle   An angle that is smaller than a right angle or 90 degrees.

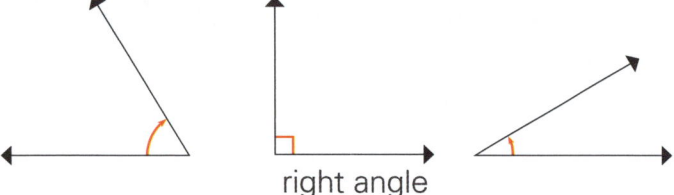

right angle

addition   The joining or adding of two numbers together to find the total. Also known as *adding, plus* and *sum*. See also *vertical addition*.

⭐⭐⭐ + ⭐⭐ = ⭐⭐⭐⭐⭐

3 and **2**   is   **5**

algorithm   A process or formula used to solve a problem in mathematics.

Examples:
horizontal algorithms
24 **+** 13 **=** 37

vertical algorithms

	T	O
	2	4
+	1	3
	3	7

analogue time   Time shown on a clock or watch face with numbers and hands to indicate the hours and minutes.

angle   The space between two lines or surfaces at the point where they meet, usually measured in degrees.

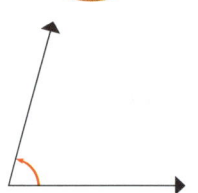

75-degree angle

anticlockwise   Moving in the opposite direction to the hands of a clock.

area   The size of an object's surface.

Example: It takes 12 tiles to cover this poster.

area model   A visual way of solving multiplication problems by constructing a rectangle with the same dimensions as the numbers you are multiplying and breaking the problem down by place value.

6 × 10 = 60
6 × 8 = 48
so
6 × 18 = 108

10   8

6

array   An arrangement of items into even columns and rows to make them easier to count.

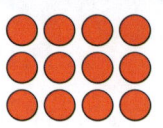

balance scale   Equipment that balances items of equal mass; used to compare the mass of different items. Also called *pan balance* or *equal arm balance*.

bar graph   A way of representing data using bars or columns to show the values of each variable.

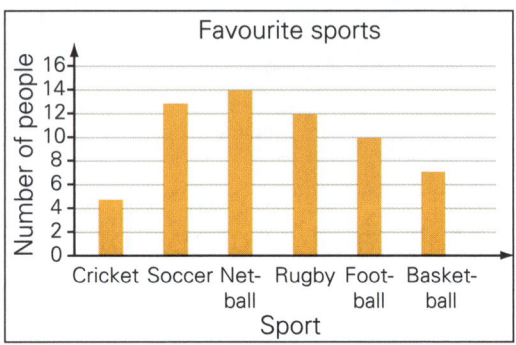

base   The bottom edge of a 2D shape or the bottom face of a 3D shape.

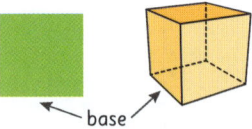

base

capacity   The amount that a container can hold.

Example: The jug has a capacity of 4 cups.

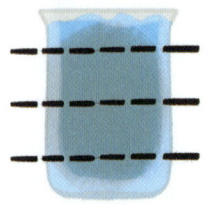

Cartesian plane   A grid system with numbered horizontal and vertical axes that allow for exact locations to be described and found.

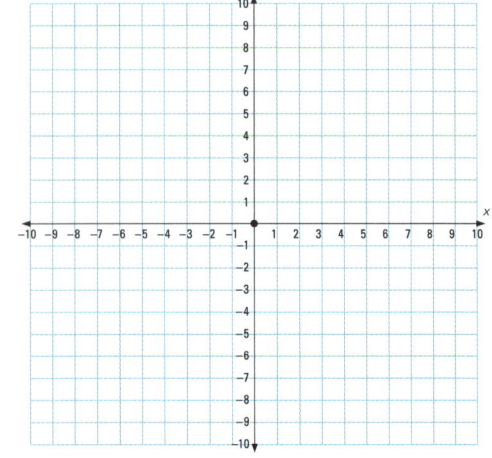

OXFORD UNIVERSITY PRESS

categorical variables   The different groups that objects or data can be sorted into based on common features.

Example: Within the category of ice-cream flavours, variables include:

vanilla   chocolate   strawberry

centimetre or *cm*   A unit for measuring the length of smaller items.

Example: Length is 80 cm.

circumference   The distance around the outside of a circle.

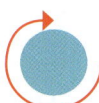

clockwise   Moving in the same direction as the hands of a clock.

common denominator   Denominators that are the same. To find a common denominator, you need to identify a multiple that two or more denominators share.

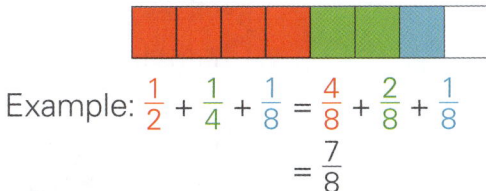

Example: $\frac{1}{2} + \frac{1}{4} + \frac{1}{8} = \frac{4}{8} + \frac{2}{8} + \frac{1}{8}$
$= \frac{7}{8}$

compensation strategy   A way of solving a problem that involves rounding a number to make it easier to work with, and then paying back or "compensating" the same amount.

Example: $24 + 99 = 24 + 100 - 1 = 123$

composite number   A number that has more than two factors, that is, a number that is not a prime number.

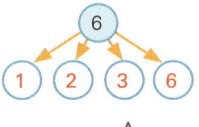

cone   A 3D shape with a circular base that tapers to a point.

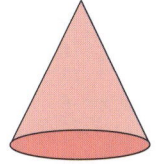

coordinates   A combination of numbers or numbers and letters that show location on a grid map.

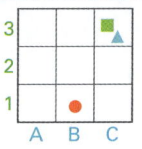

corner   The point where two edges of a shape or object meet. Also known as a *vertex*.

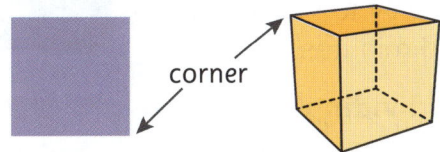

corner

cross-section   The surface or shape that results from making a straight cut through a 3D shape.

cube   A rectangular prism where all six faces are squares of equal size.

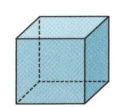

cubic centimetre or *cm³*   A unit for measuring the volume of smaller objects.

Example: This cube is exactly 1 cm long, 1 cm wide and 1 cm deep.

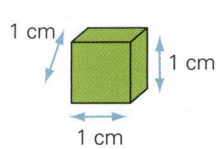

cylinder   A 3D shape with two parallel circular bases and one curved surface.

data   Information gathered through methods such as questioning, surveys or observation.

decimal fraction   A way of writing a number that separates any whole numbers from fractional parts expressed as tenths, hundredths, thousandths and so on.

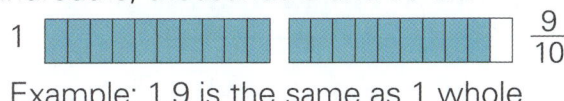

Example: 1.9 is the same as 1 whole and 9 parts out of 10 or $1\frac{9}{10}$.

degrees Celsius   A unit used to measure the temperature against the Celsius scale where 0°C is the freezing point and 100°C is the boiling point.

denominator   The bottom number in a fraction, which shows how many pieces the whole or group has been divided into.

diameter   A straight line from one side of a circle to the other, passing through the centre point.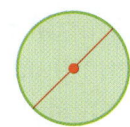

digital time   Time shown on a clock or watch face with numbers only to indicate the hours and minutes.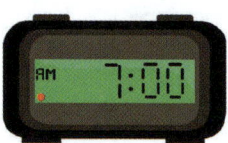

division/dividing   The process of sharing a number or group into equal parts, with or without remainders.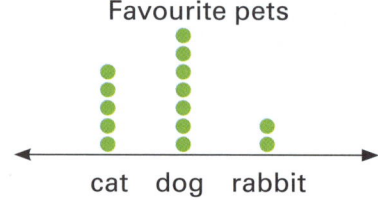

dot plot   A way of representing pieces of data using dots along a line labelled with variables.

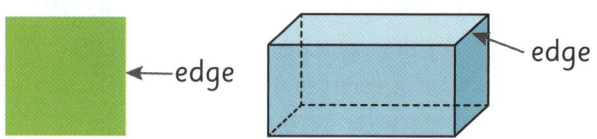
Favourite pets
cat   dog   rabbit

double/doubles   Adding two identical numbers or multiplying a number by 2.

Example: $2 + 2 = 4$ $4 \times 2 = 8$

duration   How long something lasts.

Example: Most movies have a duration of about 2 hours.

edge   The side of a shape or the line where two faces of an object meet.

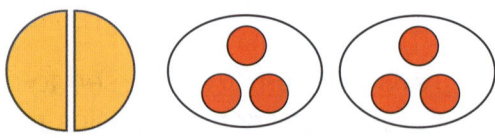
edge   edge

equal   Having the same number or value.

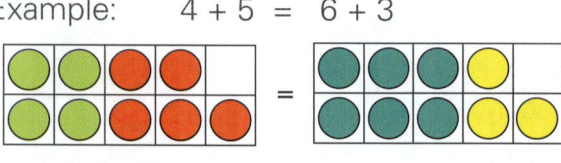

Example: Equal size   Equal numbers

equation   A written mathematical problem where both sides are equal.

Example: $4 + 5 = 6 + 3$

equilateral triangle   A triangle with three sides and angles the same size.

equivalent fractions   Different fractions that represent the same size in relation to a whole or group.

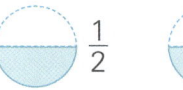 $\frac{1}{2}$ 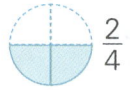 $\frac{2}{4}$  $\frac{3}{6}$  $\frac{4}{8}$

estimate   A thinking guess.

even number   A number that can be divided equally into 2.

Example: 4 and 8 are even numbers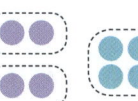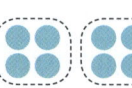

face   The flat surface of a 3D shape.

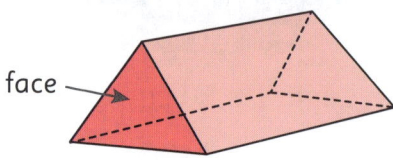
face

factor   A whole number that will divide evenly into another number.

Example: The factors of 10 are 1 and 10
2 and 5

financial plan   A plan that helps you to organise or manage your money.

flip   To turn a shape over horizontally or vertically. Also known as *reflection*.

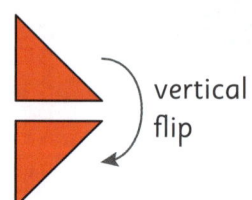
vertical flip

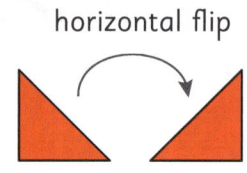
horizontal flip

fraction   An equal part of a whole or group.

Example: One out of two parts or $\frac{1}{2}$ is shaded.

grams or *g*   A unit for measuring the mass of smaller items.

1000 g is 1 kg

OXFORD UNIVERSITY PRESS

graph   A visual way to represent data or information.

 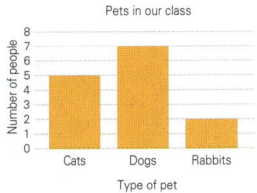

GST or Goods and Services Tax   A tax, such as 10%, that applies to most goods and services bought in many countries.

Example: Cost + GST (10%) = Amount you pay
$10 + $0.10   = $10.10

hexagon   A 2D shape with six sides.

horizontal   Parallel with the horizon or going straight across.

horizontal line

improper fraction   A fraction where the numerator is greater than the denominator, such as $\frac{3}{2}$.

integer   A whole number. Integers can be positive or negative.

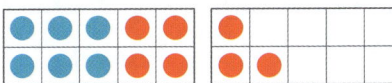

inverse operations   Operations that are the opposite or reverse of each other. Addition and subtraction are inverse operations.

Example: 6 + 7 = 13 can be reversed with
13 − 7 = 6

invoice   A written list of goods and services provided, including their cost and any GST.

Priya's Pet Store			
Tax Invoice			
Item	Quantity	Unit price	Cost
Siamese cat	1	$500	$500.00
Cat food	20	$1.50	$30.00
Total price of goods		$530.00	
GST (10%)		$53.00	
Total		$583.00	

isosceles triangle   A triangle with two sides and two angles of the same size.

jump strategy   A way to solve number problems that uses place value to "jump" along a number line by hundreds, tens and ones.

Example: 16 + 22 = 38

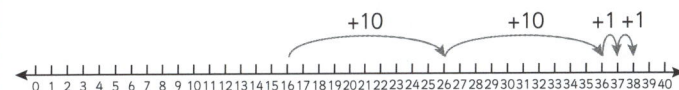

kilograms or *kg*   A unit for measuring the mass of larger items.

kilometres or *km*   A unit for measuring long distances or lengths.

kite   A four-sided shape where two pairs of adjacent sides are the same length.

legend   A key that tells you what the symbols on a map mean.

 Park  Service station  Campground Railway Road

length   The longest dimension of a shape or object.

line graph   A type of graph that joins plotted data with a line.

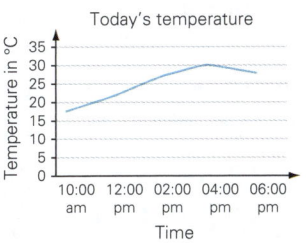

litres or L   A unit for measuring the capacity of larger containers.

Example: The capacity of this bucket is 8 litres.

mass   How heavy an object is.

Example: 4.5 kilograms   4.5 grams

metre or m   A unit for measuring the length of larger objects.

milligram or mg   A unit for measuring the mass of lighter items or to use when accuracy of measurements is important.

700 mg

millilitre or mL   A unit for measuring the capacity of smaller containers.

1000 mL is 1 litre

millimetre or mm   A unit for measuring the length of very small items or to use when accuracy of measurements is important.

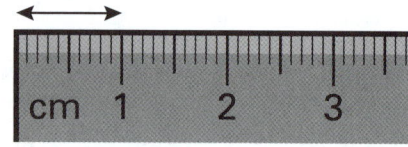

There are 10 mm in 1 cm.

mixed number   A number that contains both a whole number and a fraction.

Example: $2\frac{3}{4}$

multiple   The result of multiplying a particular whole number by another whole number.

Example: 10, 15, 20 and 100 are all multiples of 5.

near doubles   A way to add two nearly identical numbers by using known doubles facts.

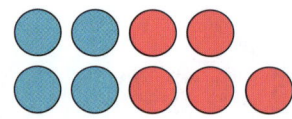

Example: $4 + 5 = 4 + 4 + 1 = 9$

net   A flat shape that when folded up makes a 3D shape.

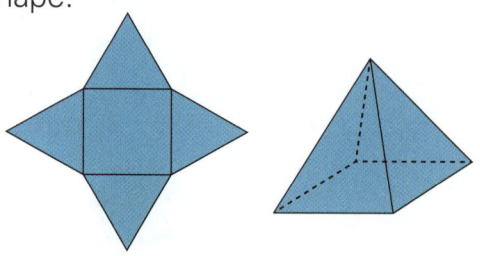

number line   A line on which numbers can be placed to show their order in our number system or to help with calculations.

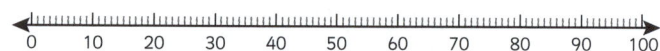

number sentence   A way to record calculations using numbers and mathematical symbols.

Example: $23 + 7 = 30$

numeral   A figure or symbol used to represent a number.

Examples: 1 – one   2 – two   3 – three

numerator   The top number in a fraction, which shows how many pieces you are dealing with.

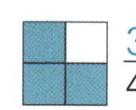

$\frac{3}{4}$

obtuse angle   An angle that is larger than a right angle or 90 degrees, but smaller than 180 degrees.

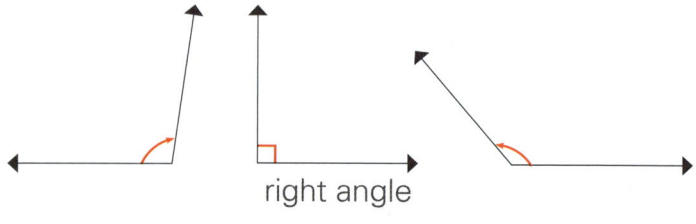

right angle

octagon   A 2D shape with eight sides.

OXFORD UNIVERSITY PRESS

odd number   A number that cannot be divided equally into 2.

Example: 5 and 9 are odd numbers.

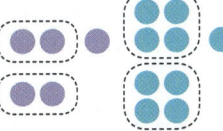

operation   A mathematical process. The four basic operations are addition, subtraction, multiplication and division.

origin   The point on a Cartesian plane where the x-axis and y-axis intersect.

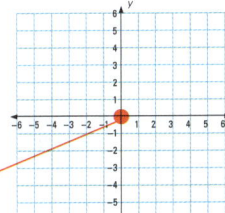

origin

outcome   The result of a chance experiment.

Example: The possible outcomes if you roll a dice are 1, 2, 3, 4, 5 or 6.

parallel lines   Straight lines that are the same distance apart and so will never cross.

parallel   parallel   not parallel

parallelogram   A four-sided shape where each pair of opposite sides is parallel.

pattern   A repeating design or sequence of numbers.

Example:
Shape pattern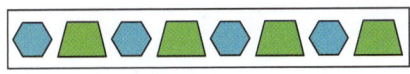
Number pattern   2, 3, 6, 8, 10, 12

pentagon   A 2D shape with five sides.

per cent or %   A fraction out of 100.

Example: $\frac{62}{100}$ or 62 out of 100

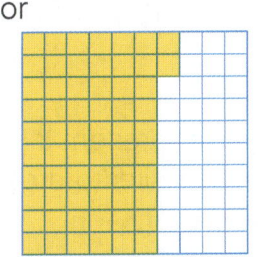

is also 62%.

perimeter   The distance around the outside of a shape or area.

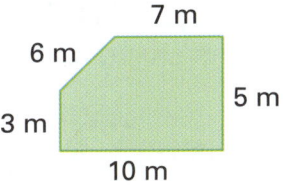

7 m
6 m
5 m
3 m
10 m

Example: Perimeter = 7 m + 5 m + 10 m + 3 m + 6 m = 31 m

pictograph   A way of representing data using pictures so that it is easy to understand.

Example: Favourite juices in our class

place value   The value of a digit depending on its place in a number.

M	H Th	T Th	Th	H	T	O
			2	7	4	8
		2	7	4	8	6
	2	7	4	8	6	3
2	7	4	8	6	3	1

polygon   A closed 2D shape with three or more straight sides.

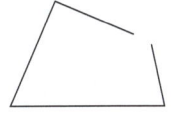

polygons   not polygons

polyhedron (plural polyhedra)   A 3D shape with flat faces.

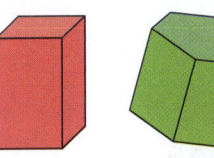
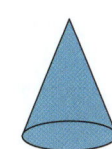

polyhedra   not polyhedra

power of   The number of times a particular number is multiplied by itself.

Example: 4^3 is 4 to the power of 3 or $4 \times 4 \times 4$.

prime number   A number that has just two factors – 1 and itself. The first four prime numbers are 2, 3, 5 and 7.

prism   A 3D shape with parallel bases of the same shape and rectangular side faces.

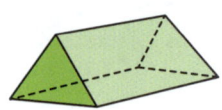 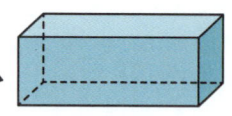 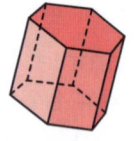

triangular prism   rectangular prism   hexagonal prism

probability   The chance or likelihood of a particular event or outcome occurring.

 Example: There is a 1 in 8 chance this spinner will land on red.

protractor   An instrument used to measure the size of angles in degrees.

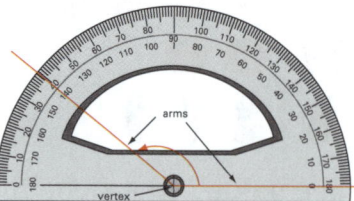

pyramid   A 3D shape with a 2D shape as a base and triangular faces meeting at a point.

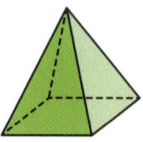 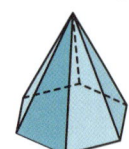

square pyramid   hexagonal pyramid

quadrant   A quarter of a circle or one of the four quarters on a Cartesian plane.

 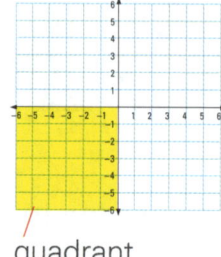

quadrant

quadrant

quadrilateral   Any 2D shape with four sides.

radius   The distance from the centre of a circle to its circumference or edge.

reflect   To turn a shape over horizontally or vertically. Also known as *flipping*.

vertical reflection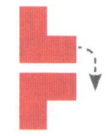

horizontal reflection

reflex angle   An angle that is between 180 and 360 degrees in size.

remainder   An amount left over after dividing one number by another.

Example: 11 ÷ 5 = 2 r1

rhombus   A 2D shape with four sides, all of the same length and opposite sides parallel.

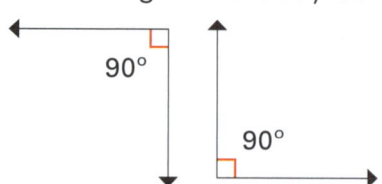

right angle   An angle of exactly 90 degrees.

90°

90°

right-angled triangle   A triangle where one angle is exactly 90 degrees.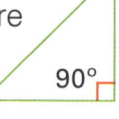

90°

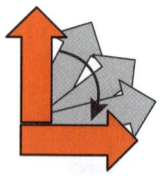 **rotate**   Turn around a point.

rotational symmetry   A shape has rotational symmetry if it fits into its own outline at least once while being turned around a fixed centre point.

1st position

Back to the start

2nd position

round/rounding   To change a number to another number that is close to it to make it easier to work with.

229 can be

rounded up to the nearest 10   OR   rounded down to the nearest 100

↑230   ↓200

scale   A way to represent large areas on maps by using ratios of smaller to larger measurements.

Example: 1 cm = 5 m

OXFORD UNIVERSITY PRESS

scalene triangle   A triangle where no sides are the same length and no angles are equal.

sector   A section of a circle bounded by two radius lines and an arc.

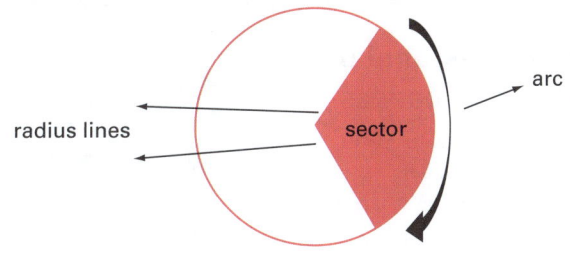

semi-circle   Half a circle, bounded by an arc and a diameter line.

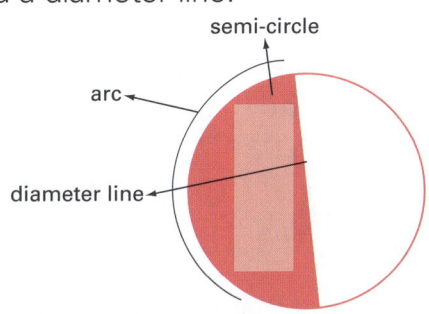

skip counting   Counting forwards or backwards by the same number each time.

Examples:
Skip counting by fives: 5, 10, 15, 20, 25, 30
Skip counting by twos: 1, 3, 5, 7, 9, 11, 13

slide   To move a shape to a new position without flipping or turning it. Also known as *translate*.

sphere   A 3D shape that is perfectly round.

split strategy   A way to solve number problems that involves splitting numbers up using place value to make them easier to work with.

Example: 21 + 14 =
20 + 10 + 1 + 4 = 35

square centimetre or c*m*²   A unit for measuring the area of smaller objects. It is exactly 1 cm long and 1 cm wide.

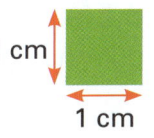

square metre or *m*²   A unit for measuring the area of larger spaces. It is exactly 1 m long and 1 m wide.

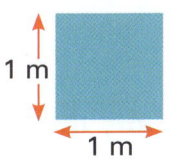

square number   The result of a number being multiplied by itself. The product can be represented as a square array.

Example: 3×3 or $3^2 = 9$

straight angle   An angle that is exactly 180 degrees in size.

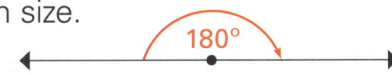

strategy   A way to solve a problem. In mathematics, you can often use more than one strategy to get the right answer.

Example: 32 + 27 = 59
Jump strategy

Split strategy
30 + 2 + 20 + 7 = 30 + 20 + 2 + 7 = 59

subtraction   The taking away of one number from another number. Also known as *subtracting, take away, difference between* and *minus*. See also *vertical subtraction*.

Example: 5 take away 2 is 3

survey   A way of collecting data or information by asking questions.

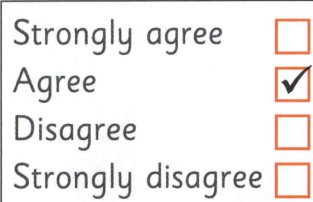

symmetry   A shape or pattern has symmetry when one side is a mirror image of the other.

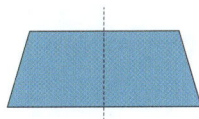

table A way to organise information that uses columns and rows.

Flavour	Number of people
Chocolate	12
Vanilla	7
Strawberry	8

tally marks A way of keeping count that uses single lines with every fifth line crossed to make a group.

term A number in a series or pattern.

Example: The sixth term in this pattern is 18.

3	6	9	12	15	18	21	24

tessellation A pattern formed by shapes that fit together without any gaps.

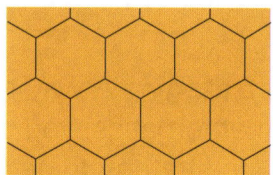

thermometer An instrument for measuring temperature.

three-dimensional or *3D*
A shape that has three dimensions – length, width and depth.
3D shapes are not flat.

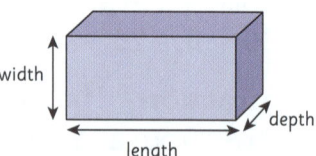

time line A visual representation of a period of time with significant events marked in.

translate To move a shape to a new position without flipping or turning it. Also known as *slide*.

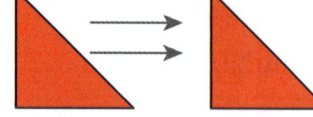

trapezium A 2D shape with four sides and only one set of parallel lines.

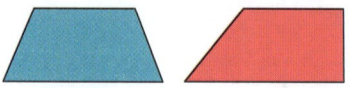

triangular number A number that can be organised into a triangular shape. The first four are:

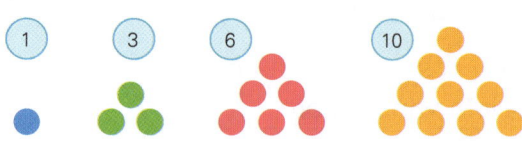

two-dimensional or *2D*
A flat shape that has two dimensions – length and width.

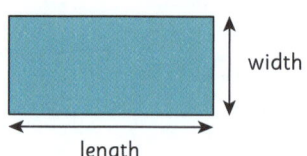

turn Rotate around a point.

unequal Not having the same size or value.

Example: Unequal size   Unequal numbers

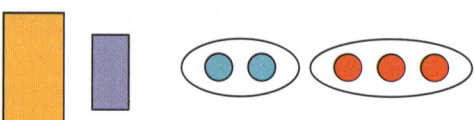

value How much something is worth.

Example:
This coin is worth 5c.   This coin is worth $1.

vertex (plural vertices) The point where two edges of a shape or object meet. Also known as a *corner*.

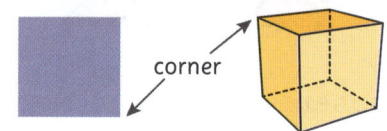

corner

vertical At a right angle to the horizon or straight up and down.

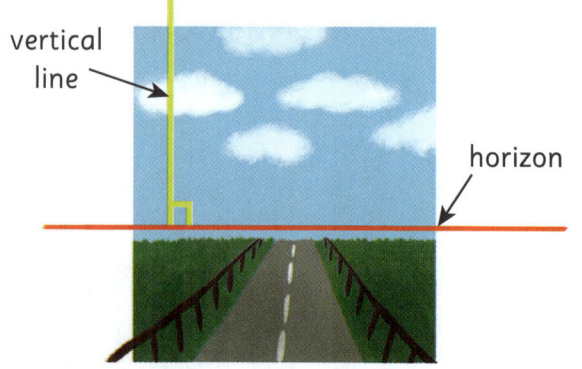

vertical line

horizon

OXFORD UNIVERSITY PRESS

vertical addition A way of recording addition so that the place-value columns are lined up vertically to make calculation easier.

	T	O
	3	6
+	2	1
	5	7

vertical subtraction A way of recording subtraction so that the place-value columns are lined up vertically to make calculation easier.

	T	O
	5	7
−	2	1
	3	6

volume How much space an object takes up.

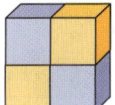

Example: This object has a volume of 4 cubes.

whole All of an item or group.

Example: A whole shape   A whole group

 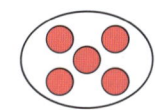

width The shortest dimension of a shape or object. Also known as *breadth*.

x-axis The horizontal reference line showing coordinates or values on a graph or map.

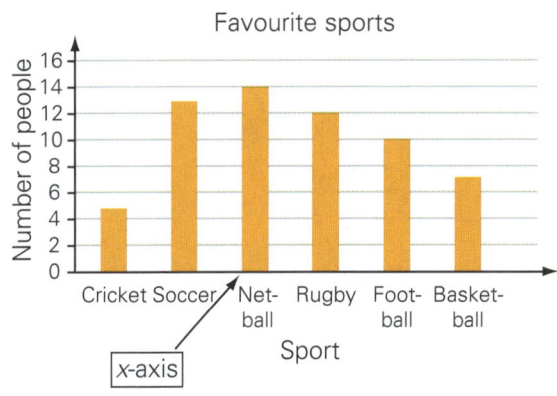

y-axis The vertical reference line showing coordinates or values on a graph or map.

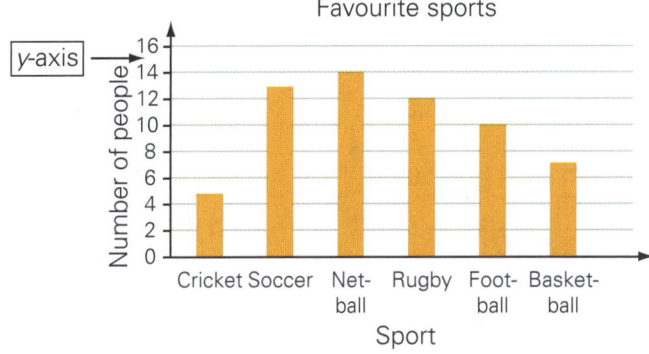

ANSWERS

Guided practice

1 a

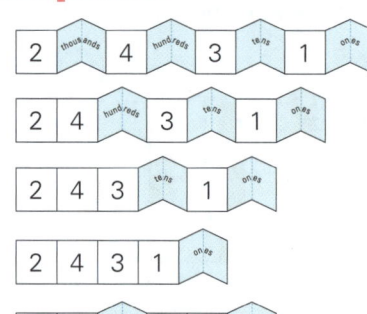

b

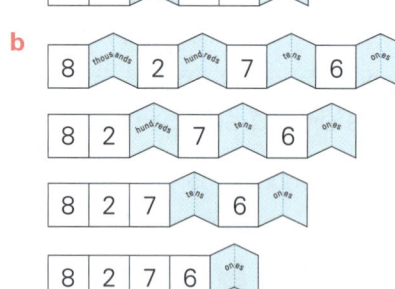

Independent practice

1 a four thousand, five hundred and sixty-eight

b eight thousand and forty-three

c seven thousand, one hundred and nine

2

Th	H	T	O
4	5	6	8
8	0	4	3
7	1	0	9

3 a 2265   **b** 3057

4

Event number	Number of people
3	5255
1	4891
5	3971
6	3812
2	1693
4	1688

5 8710

6 a 8720   **b** 8700   **c** 8730
   d 8690   **e** 8810   **f** 8610
   g 8910   **h** 8510   **i** 9710
   j 7710

7 2338

Extended practice

1 a 3790 = 3000 + 700 + 90 + 0

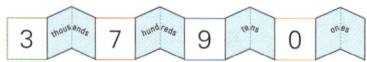

b 8052 = 8000 + 50 + 2

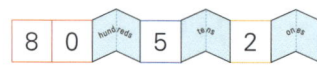

c 24 160 = 24 000 + 100 + 60

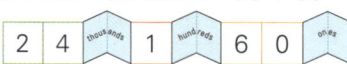

2 a 4012   **b** 6889
   c 1024   **d** 19 875

3 a 9979   1171 (or 0070)
   b 9499   1411 (or 0400)

Guided practice

1 Teacher note: The way students choose to make pairs of items will vary, however it should be apparent if the number is odd or even depending on whether or not there is a left over item.

   a odd   **b** even
   c odd   **d** even

Independent practice

1 a odd

b even

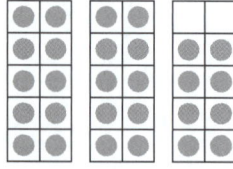

c even

d even

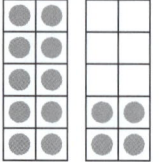

e odd

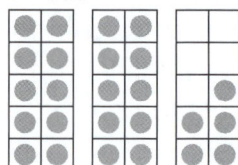

f odd

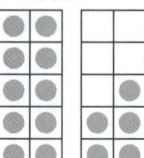

2

a

21	**23**	25	27	**29**	**31**	33	**35**	**37**

b

44	46	**48**	**50**	52	**54**	**56**	58	**60**

c

20	24	28	**32**	**36**	40	**44**	**48**	52

3 a & b:

31	32	33	34	35	36	37	38	39	40
41	42	43	44	45	46	47	48	49	50
51	52	53	54	55	56	57	58	59	60

   c 2, 4, 6, 8, 0 (in any order)

   d 1, 3, 5, 7, 9 (in any order)

4

Odd	Even
143	76
103	258
575	1974
1361	3870
867	5002
9999	9998

5 a odd   **b** even
   c even   **d** even

Extended practice

1 a 8   **b** 24   **c** 36   **d** even
2 a 8   **b** 28   **c** 30   **d** even
3 a 9   **b** 27   **c** 39   **d** odd
4 a 11   **b** 27   **c** 37   **d** odd
5 a even   **b** odd   **c** odd
   d even   **e** even   **f** odd

Guided practice

1 a 7 and 17   **b** 8 and 18
   c 10 and 20   **d** 5 and 25

OXFORD UNIVERSITY PRESS

Independent practice

1 **a** 9 and 29   **b** 8 and 18
   c 6 and 26   **d** 9 and 39
   e 10 and 40

2 **a** 60   **b** 8, 80
   c 10, 100   **d** 4, 20, 20
   e If 8 + 8 = 16,
     then 80 + 80 = 160.
   f If 1 + 1 = 2,
     then 100 + 100 = 200.
   g If 6 + 6 = 12,
     then 600 + 600 = 1200.
   h If 7 + 7 = 14,
     then 700 + 700 = 1400.

3 **a** 23 + 12 = 30 + 5 = 35
   b 50 + 7 = 57   **c** 80 + 7 = 87
   d 80 + 9 = 89   **e** 60 + 10 = 70

4 **a** 6 + 4 + 7 = 17
   b 25 + 5 + 4 = 34
   c 17 + 3 + 2 + 4 = 26
   d 11 + 19 + 3 + 2 = 35

5 **a** 180   **b** 98   **c** 41
   d 40   **e** 89   **f** 1000
   g 78   **h** 50

Extended practice

1 **a** 12 + 8 + 7 = 27
   b 23 + 7 + 12 = 42
   c 221 + 39 + 8 = 268

2 **a** 54 + 39 = 93
   b 221 + 23 = 244
   c 135 + 54 = 189
   d 221 + 135 = 356

UNIT 1: Topic 4

Guided practice

1 37

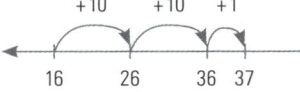

2 59

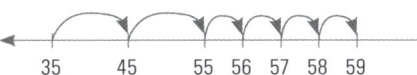

3 179

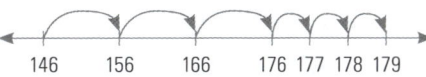

Independent practice

1 **a** 97

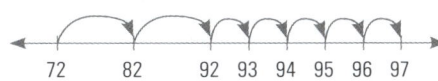

   b 169

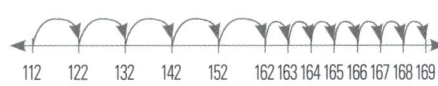

   c 294

   d 361

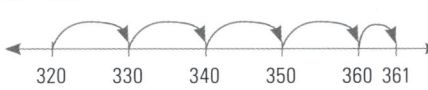

   e 439

Guided practice

1 **a** 96   **b** 168   **c** 387
   d 746   **e** 879   **f** 996
   g 474   **h** 888   **i** 909

Independent practice

1 **a**

	2	8
+	3	1
	5	9

   b

	6	3
+	3	5
	9	8

   c

	4	6
+	2	2
	6	8

   d

	3	5	8
+	4	2	1
	7	7	9

   e

	4	8	0
+	2	1	7
	6	9	7

   f

	8	9	1
+	2	0	6
1	0	9	7

2 **a**

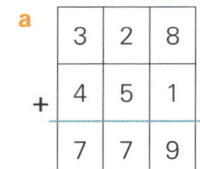

	3	2	8
+	4	5	1
	7	7	9

   b

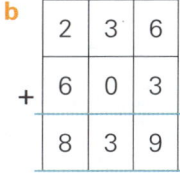

	2	3	6
+	6	0	3
	8	3	9

Extended practice

1 **a** 802

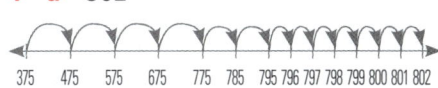

   b 923

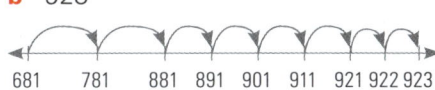

2 **a** 1788   **b** 3519   **c** 7587

3 867   Teacher to check strategies.
Teacher: Look for students
who choose an appropriate
strategy, and can follow the steps
sequentially to find the correct
answer.

UNIT 1: Topic 5

Guided practice

1 **a** 3, 13   **b** 7, 17   **c** 2, 12
   d 4, 24

Independent practice

1 **a** 2, 12   **b** 1, 21   **c** 5, 15
   d 6, 26   **e** 3, 33   **f** 2, 82
   g 3, 93

2 **a** 35 − 13 = 35 − 10 − 3 = 22
   b 48 − 15 = 48 − 10 − 5 = 33
   c 52 − 21 = 52 − 20 − 1 = 31
   d 67 − 34 = 67 − 30 − 4 = 33
   e 96 − 25 = 96 − 20 − 5 = 71
   f 124 − 13 = 124 − 10 − 3 = 111
   g 389 − 57 = 389 − 50 − 7 = 332

3 **a** 26 − 8 = 26 − 6 − 2 = 18
   b 32 − 7 = 32 − 2 − 5 = 25
   c 35 − 9 = 35 − 5 − 4 = 26
   d 21 − 6 = 21 − 1 − 5 = 15
   e 43 − 5 = 43 − 3 − 2 = 38
   f 64 − 7 = 64 − 4 − 3 = 57
   g 76 − 9 = 76 − 6 − 3 = 67
   h 145 − 8 = 145 − 5 − 3 = 137

Extended practice

1 **a** 2, 20   **b** 7, 70   **c** 4, 40
   d 2, 200   **e** 1, 100

2 **a** 14   **b** 59   **c** 141   **d** 124

Teacher: Look for students who
can articulate how they arrived
at the answer and what mental
strategies they used.

UNIT 1: Topic 6

Guided practice

1 23

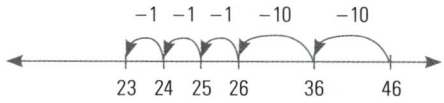

2 23

3 222

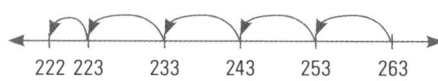

Independent practice

1 **a** 64

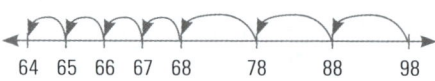

b 317

c 747

d 473

e 169

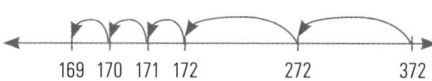

Guided practice

1 **a** 23 **b** 447 **c** 575
d 732 **e** 223 **f** 504
g 200 **h** 730 **i** 333

Independent practice

1 **a**

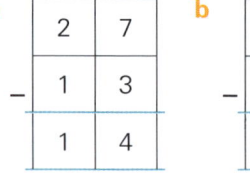

	2	7
−	1	3
	1	4

b

	5	3
−	3	1
	2	2

c

8	6
−	3	6
	5	0

d

1	7	3
−	1	6	2
	0	1	1

e

7	9	7
−	4	9	3
	3	0	4

f

8	9	1
−	2	0	6
	6	8	5

2 **a**

9	8
−	5	7
	4	1

b

6	4	5
−	4	1	4
	2	3	1

Extended practice

1 **a** 526

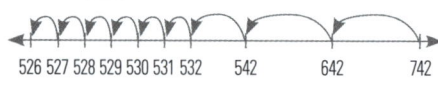

b 285

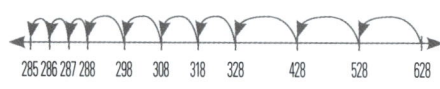

2 **a** 5214 **b** 2662 **c** 2511

3 515 Teacher to check strategy. Teacher: Look for students who choose an appropriate strategy and can follow the steps sequentially to find the correct answer.

UNIT 1: Topic 7

Guided practice

1 **a** 7 **b** 24 **c** 38
2 **a** 9 **b** 27 **c** 43

Independent practice

1 **a** 6 + 4 = 10, 4 + 6 = 10,
10 − 6 = 4, 10 − 4 = 6

b 17 + 7 = 24, 7 + 17 = 24,
24 − 7 = 17, 24 − 17 = 7

c 17 + 12 = 29, 12 + 17 = 29,
29 − 17 = 12, 29 − 12 = 17

d 40 + 8 = 48, 8 + 40 = 48,
48 − 8 = 40, 48 − 40 = 8

e 45 + 37 = 82, 37 + 45 = 82,
82 − 37 = 45, 82 − 45 = 37

f 100 + 26 = 126, 26 + 100 = 126,
126 − 26 = 100, 126 − 100 = 26

2 **a** 14 + 17 = 31, 17 + 14 = 31,
31 − 14 = 17, 31 − 17 = 14

b 32 + 46 = 78, 46 + 32 = 78,
78 − 32 = 46, 78 − 46 = 32

c 15 + 33 = 48, 33 + 15 = 48,
48 − 15 = 33, 48 − 33 = 15

d 16 + 39 = 55, 39 + 16 = 55,
55 − 16 = 39, 55 − 39 = 16

e 97 + 70 = 167, 70 + 97 = 167,
167 − 97 = 70, 167 − 70 = 97

f 143 + 135 = 278,
135 + 143 = 278,
278 − 143 = 135,
278 − 135 = 143

Extended practice

1 **a** 34 + 28 is the same as
34 + 30 − 2 = 62

b 26 + 29 is the same as
26 + 30 − 1 = 55

c 53 + 49 is the same as
53 + 50 − 1 = 102

d 45 + 27 is the same as
45 + 30 − 3 = 72

e 54 + 17 is the same as
54 + 20 − 3 = 71

2 **a** 2 × 10 = 20, 10 × 2 = 20,
20 ÷ 2 = 10, 20 ÷ 10 = 2

b 4 × 12 = 48, 12 × 4 = 48,
48 ÷ 4 = 12, 48 ÷ 12 = 4

c 8 × 7 = 56, 7 × 8 = 56,
56 ÷ 7 = 8, 56 ÷ 8 = 7

d 9 × 11 = 99, 11 × 9 = 99,
99 ÷ 11 = 9, 99 ÷ 9 = 11

3 **a** 73 **b** 1532

UNIT 1: Topic 8

Guided practice

1 **a** 15 shared between 3 is 5
b 12 shared between 6 is 2
c 28 shared between 4 is 7

2 **a** 3 groups of 3 = 9
b 8 groups of 2 = 16
c 3 groups of 6 = 18

Independent practice

1 **a** 3 × 4 = 12, 4 × 3 = 12
b 5 × 10 = 50, 10 × 5 = 50
c 5 × 6 = 30, 6 × 5 = 30
d 4 × 10 = 40, 10 × 4 = 40

2 Note: Answers can be in any order.

a 3 × 9 = 27, 9 × 3 = 27,
27 ÷ 3 = 9, 27 ÷ 9 = 3

b 10 × 2 = 20, 2 × 10 = 20,
20 ÷ 2 = 10, 20 ÷ 10 = 2

c 8 × 5 = 40, 5 × 8 = 40,
40 ÷ 5 = 8, 40 ÷ 8 = 5

OXFORD UNIVERSITY PRESS

d $7 \times 10 = 70$, $10 \times 7 = 70$,
$70 \div 10 = 7$, $70 \div 7 = 10$

3
a 3
b 6
c 9
d 12
e 15
f 18
g 21
h 24
i 27
j 30

4
a $3 \div 3 = 1$
b $6 \div 3 = 2$
c $9 \div 3 = 3$
d $12 \div 3 = 4$
e $15 \div 3 = 5$
f $18 \div 3 = 6$
g $21 \div 3 = 7$
h $24 \div 3 = 8$
i $27 \div 3 = 9$
j $30 \div 3 = 10$

5
a 4 **b** 9 **c** 6
d 7 **e** 7 **f** 9

6
a $5 \times 4 = 20$ or $4 \times 5 = 20$
b $9 \times 2 = 18$ or $2 \times 9 = 18$
c $6 \times 10 = 60$ or $10 \times 6 = 60$
d $7 \times 5 = 35$ or $5 \times 7 = 35$
e $2 \times 7 = 14$ or $7 \times 2 = 14$
f $9 \times 10 = 90$ or $10 \times 9 = 90$

Extended practice

1 a 15 **b** 30 **c** 35 **d** 50
2 a 8 **b** 4 **c** 3 **d** 12
3 a

Name	Number of items sold	Cost per item	Amount raised
Mika	8	$5	$40
Andy	10	$2	$20
Serena	6	$10	$60
Sophia	5	$9	$45
Hao	9	$4	$36

b Andy **c** Serena **d** $80
e $100 **f** 7

UNIT 1: Topic 9

Guided practice

1
a 3, 6, 9, 12, 15, 18
b 2, 4, 6, 8, 10, 12, 14, 16
c 10, 20, 30
d 5, 10, 15, 20, 25, 30, 35
e 3, 6, 9, 12, 15, 18, 21, 24

Independent practice

1
a $8 \times 4 = 8 \times 2 \times 2 = 16 \times 2 = 32$
b $20 \times 4 = 20 \times 2 \times 2 = 40 \times 2$
$= 80$
c $12 \times 4 = 12 \times 2 \times 2 = 24 \times 2$
$= 48$
d $30 \times 4 = 30 \times 2 \times 2 = 60 \times 2$
$= 120$

2
a $16 \div 2 = 8$, $8 \div 2 = 4$,
so $16 \div 4 = 4$
b $40 \div 2 = 20$, $20 \div 2 = 10$,
so $40 \div 4 = 10$
c $60 \div 2 = 30$, $30 \div 2 = 15$,
so $60 \div 4 = 15$

3
a $2 \times 13 = 26$, so $26 \div 2 = 13$
b $3 \times 9 = 27$, so $27 \div 3 = 9$
c $5 \times 9 = 45$, so $45 \div 5 = 9$
d $5 \times 11 = 55$, so $55 \div 5 = 11$
e $10 \times 12 = 120$, so $120 \div 10 = 12$

4
a 30 **b** 90 **c** 10 **d** 3
e 32 **f** 12 **g** 6 **h** $80

Extended practice

1 a 64 **b** 21 **c** 60 **d** 20

UNIT 1: Topic 10

Guided practice

1
a $2 \times 20 + 2 \times 6 = 40 + 12 = 52$
b $4 \times 10 + 4 \times 4 = 40 + 16 = 56$
c $3 \times 10 + 3 \times 9 = 30 + 27 = 57$

Independent practice

1
a $5 \times 13 = 5 \times 10 + 5 \times 3 =$
$50 + 15 = 65$
b $6 \times 21 = 6 \times 20 + 6 \times 1 =$
$120 + 6 = 126$
c $4 \times 32 = 4 \times 30 + 4 \times 2 =$
$120 + 8 = 128$
d $7 \times 24 = 7 \times 20 + 7 \times 4 =$
$140 + 28 = 168$
e $5 \times 45 = 5 \times 40 + 5 \times 5 =$
$200 + 25 = 225$
f $8 \times 33 = 8 \times 30 + 8 \times 3 =$
$240 + 24 = 264$
g $3 \times 58 = 3 \times 50 + 3 \times 8 =$
$150 + 24 = 174$

2

a 108

×	20	7
4	80	28

b 216

×	30	6
6	180	36

c 265

×	50	3
5	250	15

d 186

×	60	2
3	180	6

e 420

×	80	4
5	400	20

f 192

×	40	8
4	160	32

g 190

×	90	5
2	180	10

Extended practice

1 Teacher: Look for students who are able to successfully interpret the problems and choose an appropriate strategy to solve each problem. Students also need to be able to accurately apply the strategy to find the correct answer.
a 148 **b** 96 **c** 190
d $4 \times 26 = 104 - 3 = 101$

Unit 1: Topic 11

Guided practice

1 a 18 **b** 30
c 14 **d** 40

Independent practice

1 a 28 **b** 38
c 26 **d** 32

2 Look for students who link numbers that add to 10 or multiples of 10. Likely answers are listed below.
a $6 + 4 + 7 + 3 = 20$
b $18 + 2 + 5 + 5 = 30$
c $14 + 6 + 9 + 1 = 30$
d $23 + 7 + 6 + 14 = 50$

3 Look for students who group numbers that are easy to multiply. Possible answers are listed below.
a $5 \times 2 = 10 \times 7 = 70$
b $6 \times 6 = 36$
c $5 \times 2 = 10 \times 3 = 30$
d $2 \times 3 = 6 \times 7 = 42$

4
a 23 [$23 - 9 = 14$]
b 11 [$11 + 14 = 25$]
c 27 [$27 \div 3 = 9$]
d 8 [$8 \times 5 = 40$]
e 21 [$21 + 21 = 42$]
f 55 [$55 \div 5 = 11$]
g 67 [$67 - 24 = 43$]
h 12 [$12 \times 10 = 120$]

5 Teachers could ask students to explain their strategies and/or share them with their peers. Possible solutions are listed below.

  a $5 \times 3 = 15$

  b $3 + 17 = 20 + 4 = 24$

  c $2 \times 5 = 10 \times 9 = 90$

  d $18 + 12 = 30 + 10 = 40$

  e $6 \times 4 = 24$

  f 9 [inverse operation of question 3]

  g Link: $[3 + 7] + [16 + 14] + [8 + 2] = 10 + 30 + 10 = 50$

  h $5 \times 7 + 1 = 36$

Extended practice

1 Look for students who use some or all of the strategies from this topic.

  a Tran is incorrect: $15 \times 10 = 10 \times 15$. Both have 150 cards.

  b Look for students who link numbers that add to 10 or multiples of 10. The total is $80 [3 + 7 + 8 + 12 + 4 + 16 + 11 + 9 + 5 + 5].

  c 64 ($7 \times 9 + 1$)

  d 22 books ($110 \div 5$)

UNIT 2: Topic 1

Guided practice

1 a Three of the five parts should be shaded.

b One of the three parts should be shaded.

c One of the two parts should be shaded.

d Three of the four parts should be shaded.

e Four of the five parts should be shaded.

f Two of the three parts should be shaded.

Independent practice

1 a $\frac{1}{3}$ **b** $\frac{3}{8}$ **c** $\frac{2}{5}$
  d $\frac{2}{4}$ or $\frac{1}{2}$ **e** $\frac{5}{8}$ **f** $\frac{4}{5}$
  g $\frac{3}{4}$ **h** $\frac{3}{4}$
  i $\frac{4}{8}$ or $\frac{1}{2}$ **j** $\frac{4}{6}$

2

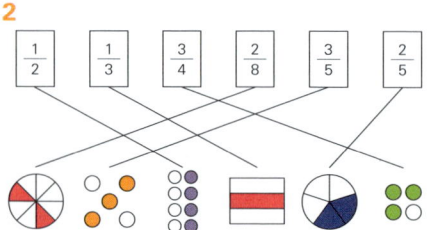

---

3 a–d Teacher to check. Teacher: Look for students who can divide the shapes into the correct number of parts and who show an understanding of the need to make the parts equal in size.

4 a fifths   **b** halves
  c fifths   **d** halves

Extended practice

1 a, c & e Teacher to check. Teacher: Look for students who can draw lines to divide the square into the correct number of parts and who show an understanding that fractions are made up of parts of equal size.

  b $\frac{1}{2}$ or a half

  d $\frac{1}{4}$ or a quarter   **f** $\frac{1}{8}$

  g Any 5 of the parts may be coloured in.

  h $\frac{5}{8}$ **i** $\frac{3}{8}$

2 $\frac{1}{8}, \frac{1}{4}, \frac{1}{2}, \frac{5}{8}$

UNIT 2: Topic 2

Guided practice

1 a

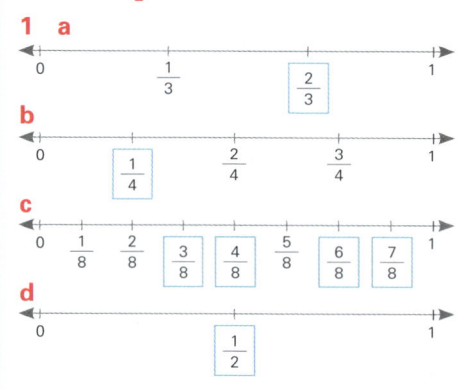

Independent practice

1 a

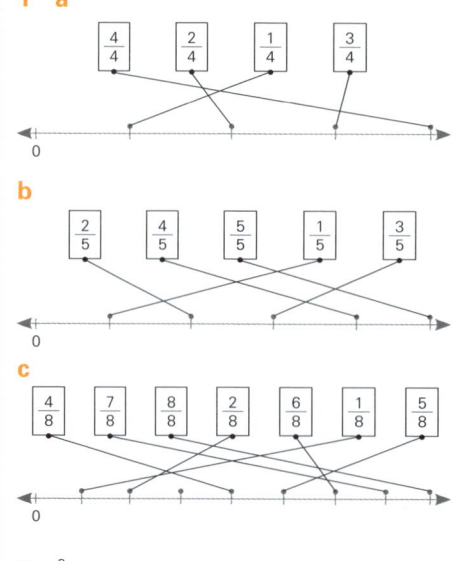

2 $\frac{3}{8}$

---

3 a 8   **b** 2   **c** 5
  d 3   **e** 4

4 a $\frac{1}{2}$ **b** $\frac{1}{5}$ **c** $\frac{1}{3}$
  d $\frac{2}{4}$ **e** $\frac{2}{3}$ **f** $\frac{4}{5}$

5 Teacher to check. Teacher: Look for students who can articulate that both fractions represent a whole (or one) and are therefore equal.

Extended practice

1 a

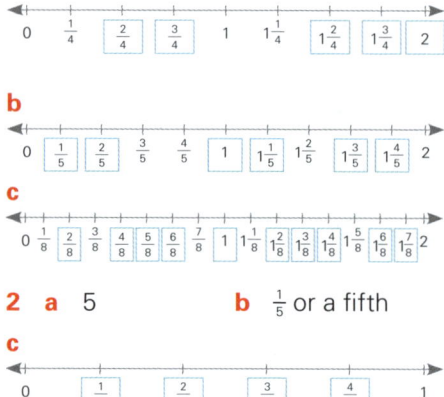

b

c

2 a 5   **b** $\frac{1}{5}$ or a fifth

c

d $1\frac{1}{5}$ **e** $\frac{5}{5}$

UNIT 3: Topic 1

Guided practice

1 Teacher to check. Teacher: Look for students who demonstrate an understanding of the value of coins and who show fluency in their addition skills.

  Some possible combinations include:

a 50c and 20c, three 20c and one 10c, or one 50c and two 10c coins

b one $1 coin, two 50c coins or five 20c coins

c two 20c coins, four 10c coins, or three 10c coins and two 5c coins

Independent practice

1 Teacher to check. Teacher: Look for students who demonstrate fluency with coins and calculations by making the given total using only three coins. Likely answers are:

a three 10c coins

b one 50c and two 20c coins

c one dollar coin and two 10c coins or two 50c coins and one 20c coin

d one $2 coin and two 5c coins or two $1 coins and one 10c coin

2 Students may choose to draw or write answers.

a one $1 coin and one 20c coin

b one 50c, 10c and 5c coin

OXFORD UNIVERSITY PRESS

c two $2, one $1 and one 10c coin

d one $1, one 50c, two 20c coins and one 5c coin

e one $2 and one 50c coin

f one 50c, one 20c, one 10c and one 5c coin

3 Students may choose to draw or write answers.

   a $1.50   **b** $3.25
   c $4.10   **d** $2.95

4 **a** 80c   **b** 40c
   c 35c   **d** 30c

Extended practice

1 **a** 20c   **b** 70c   **c** 45c
   d $1.05   **e** $1.80   **f** $3.00

2 **a** $7.60

b Teacher to check. Teacher: Look for students who can demonstrate proficiency with money calculations by accurately reaching the total.

c She would not receive any change. Teacher: Look for students who are able to make the connection that 2c cannot be given as change, and the amount would therefore need to be rounded up to $7.60.

UNIT 4: Topic 1

Guided practice

1 **a**

| 3 | 8 | 13 | **18** | **23** | **28** | **33** | **38** | **43** | **48** | **53** |

b

| 54 | 51 | 48 | **45** | **42** | **39** | **36** | **33** | **30** | **27** | **24** |

c

| 6 | 12 | **18** | **24** | **30** | **36** | **42** | **48** | **54** | **60** |

d

| 65 | 61 | 57 | **53** | **49** | **45** | **41** | **37** | **33** | **29** | **25** |

e

| 24 | 34 | 44 | **54** | **64** | **74** | **84** | **94** | **104** | **114** | **124** |

Independent practice

1 **a** Add 10   **b** Subtract 5
   c Add 7

2 **a** Subtract 4

In	Out
52	48
36	32
44	40
28	24

---

b Subtract 2

In	Out
13	11
31	29
5	3
47	45

c Add 8

In	Out
19	27
44	52
62	70
53	61

d Subtract 9

In	Out
64	55
48	39
56	47
30	21

3 **a**

•	••	•••	••••	•••••	••••••
1	3	5	7	9	11

b Add 2

4 **a**

⠿⠿					
18	15	12	9	6	3

b Subtract 3

5 **a & b** Teacher to check. Teacher: Look for students who can create correct addition and subtraction patterns, and whose rules match their patterns.

Extended practice

1 **a** Add 5, subtract 1
   b Subtract 2, add 3

2 **a**

| 1 | 2 | 5 | 6 | **9** | **10** | **13** | **14** | **17** | **18** |

b

| 56 | 54 | 51 | 49 | **46** | **44** | **41** | **39** | **36** | **34** |

3 Teacher to check. Teacher: Look for students who can identify the two steps in their pattern and correctly use their rule to complete the numbers in the pattern.

UNIT 4: Topic 2

Guided practice

1

a 7 + ⬚5 = 12

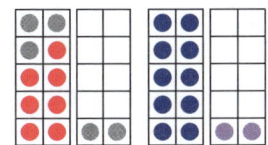

b 19 − ⬚4 = 15

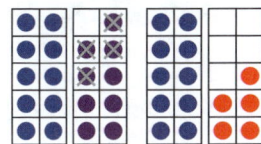

c 10 + ⬚8 = 18

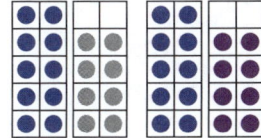

d 16 − ⬚7 = 9

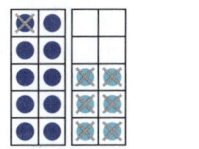

e 17 = ⬚3 + 14

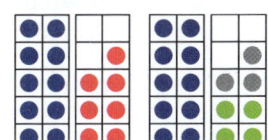

f 16 = 19 − ⬚3

Independent practice

1 **a** 4   **b** 2   **c** 8   **d** 15
   e 20   **f** 20

2 **a** +   **b** +   **c** −   **d** +
   e −   **f** −   **g** −   **h** +

3 Teacher: Students may use different strategies resulting in number sentences different from those below. Accept reasonable responses that result in the correct answers. The most likely are:

   a 46 + 19 = 65

   b 84 − 32 = 52

   c $74 − $49 = $25 or $49 + $25 = $74

   d 42 + 14 + 28 = 84

e 200 − 153 = 47 or
153 + 47 = 200

f 100 − 32 − 41 = 27 or
32 + 41 = 73, 100 − 73 = 27

Extended practice

1 a 295 **b** Daina **c** 34
d Tanmay and Jonas **e** 789
f 42

2 a False **b** True **c** True
d True **e** True

UNIT 5: Topic 1

Guided practice

1 a 5 cm **b** 15 cm **c** 3 cm
d 10 cm

2 a the pencil **b** the paper clip
c the matchstick

Independent practice

1 a–b Teacher to check. Teacher:
Look for students who can make
reasonable estimates in both cm
and m and who can accurately
measure their chosen items.

2 Teacher: The most likely answers
are shown here. Accept other
answers if students can justify
their choices – e.g. "I would use
cm to measure the basketball
court because it has to be an exact
length."

a m **b** cm **c** m **d** m
e cm **f** cm

3 a 13 cm **b** 5 m
c 12 cm **d** 6 m

Guided practice

1 a 4 cm² **b** 12 cm² **c** 8 cm²
d 8 cm² **e** 2 cm² **f** 6 cm²

2 a b **b** e

3 c and **d**

Independent practice

1 a–d Teacher to check.
Teacher: Look for students who can
accurately make the shapes based
on the specifications and who show
an awareness of the basic concept
of area – e.g. the squares that make
up each shape must have at least
one joining edge.

2 47 cm²

3 a Teacher to check.
b 36 cm² **c** 6 cm² **d** 42 cm²

Extended practice

1 Teacher: Given that millimetres are
a very small unit of measurement,

answers 1 or 2 mm either side of
those given here are acceptable.

a 45 mm **b** 31 mm **c** 6 mm
d 10 mm **e** 22 mm **f** 17 mm

2 a 6 m² **b** 15 m² **c** 2 m²
d 9 m²

3 4 m²

UNIT 5: Topic 2

Guided practice

1 a 4 cubic centimetres or 4 cm³
b 5 cubic centimetres or 5 cm³
c 11 cubic centimetres or 11 cm³
d 9 cubic centimetres or 12 cm³
e 12 cubic centimetres or 12 cm³
f 6 cubic centimetres or 6 cm³

Independent practice

1 a 2 **b** 6 **c** 12 cm³
2 a 3 **b** 4 **c** 12 cm³
3 a 3 **b** 8 **c** 24 cm³
4 a green **b** blue and pink
c 12 cm³

Guided practice

1 a B E **b** A F G
c C D **d** F **e** B

Independent practice

1 a A and E
b E and G, D and G or D and E
c 680 mL
d 1100 mL or 1 L and 100 mL

2 a–c Teacher to check. Teacher:
Look for students who can make
sound estimations of capacity
in relation to a litre, and who are
then able to accurately measure
to check whether each container
holds more or less than 1 litre.

Extended practice

1 & 2 Teacher to check. Teacher:
Look for students who
demonstrate an understanding
of the concept of volume by
being able to create an object
that meets the given criterion.
Drawing the objects may be
challenging, and this may be a
useful discussion point with
the class.

3 a & b Teacher to check. Teacher:
Look for students who show an
understanding of millilitres as a
unit of capacity by making close
estimates for their containers.

Students should also be able
to use instruments such as
measuring jugs to check the exact
measurement.

c & d Teacher: Answers will
vary depending on students
responses to a & b. Look for
students who demonstrate
an understanding of capacity
by correctly identifying items
with the largest and smallest
capacities.

UNIT 5: Topic 3

Guided practice

1 a C E A B F D
b E A B C F D

2 a the elephant **b** the 20c coin

Independent practice

1 a & b Teacher to check. Teacher:
Look for students who can make
reasonable estimates about the
mass of items relative to 1 kg, and
who can use the language of mass
to justify their reasoning.

c Teacher: Responses will depend
on items chosen by students.
Look for students who can use a
pan balance to check the mass of
their objects.

d Teacher to check. Teacher: Look
for students who are able to make
reasonable estimates of objects
that might have a mass of 1 kg
and who can correctly use a pan
balance with a 1 kg weight to
check their estimates.

2 a & b Teacher to check. Teacher:
Look for students who can make
reasonable estimates of items
with a mass of less than 500 g,
and who can use a pan balance to
find the mass of their items.

2 c & d Teacher to check. Look for
students who make reasonable
estimates of objects that might
have a mass of 500 g and who
are able to accurately check their
estimates using a pan balance.

3 a–c Teacher: Responses will vary
depending on the objects chosen
by students. Look for students who
demonstrate an understanding of
the concept of balance in mass and
who can use their initial estimate
to refine their judgement of the
number of items likely to balance
the subsequent weights.

4 a 2 **b** 5 **c** 10 **d** 20
e 15 **f** 25

OXFORD UNIVERSITY PRESS

Extended practice

1 **a** 2 kg **b** 4 kg **c** 200 g
  d 500 g

2 **a** 250 g **b** 2 kg **c** 20 g
  d 125 g

3 **a** 2 kg **b** 300 g
  c $1\frac{1}{2}$ kilograms or 1 kg and 500 g
  d $3\frac{1}{2}$ kilograms or 3 kg and 500 g

UNIT 5: Topic 4

Guided practice

1 **a** 10 past 8, 8:10
  b 20 to 5, 4:40
  c half (or 30) past 1, 1:30
  d 9 to 10, 9:51
  e 17 to 7, 6:43
  f 19 past 11, 11:19

Independent practice

1 **a**  **b**

c  **d**

e  **f**

2 **a**

  b

  c

3 **a**  **b**

c  **d**

e  **f**

4 **a** 10 minutes **b** 5 minutes
  c 20 minutes
  d 60 minutes or 1 hour

5 **a** 60 **b** 120 **c** 30 **d** 90
  e 15 **f** 45

6 **a** 60 **b** 120 **c** 300 **d** 600
  e 210 **f** 630

Extended practice

1 **a**

b 7:57 **c** 3 minutes to 8

2 **a**

b 5:22 **c** 22 minutes past 5

3 **a** 4 **b** 32 **c** 60 **d** 44

4 **a** 3 minutes
  b 1 hour and 18 minutes OR
    78 minutes
  c 11 hours and 58 minutes

UNIT 6: Topic 1

Guided practice

1

rectangle
parallelogram
rhombus
kite
trapezium

- regular shape
- type of parallelogram

- irregular
- 2 pairs of adjacent sides the same length

- irregular
- 1 pair of parallel sides

- irregular
- 2 pairs of parallel sides

- irregular
- 4 right angles
- 2 pairs of parallel sides

Independent practice

1 Teacher: In many cases, there are multiple answers for the name of a shape – e.g. a square could also be known as a rectangle or a quadrilateral. The most likely responses are given below; however, accept any correct response.

a hexagon Parallel lines: yes
  Regular: yes No. of sides: 6

b rhombus Parallel lines: yes
  Regular: yes No. of sides: 4

c pentagon Parallel lines: no
  Regular: no No. of sides: 5

d hexagon Parallel lines: yes
  Regular: no No. of sides: 6

e triangle Parallel lines: no
  Regular: yes No. of sides: 3

2 Teacher: As with question 1, students' descriptions may vary.

a pentagon, 5 sides, all sides equal, no parallel sides

b trapezium, 4 sides, type of quadrilateral, 1 pair of parallel sides

c triangle, 1 right angle, no sides equal, no parallel sides

d octagon, 8 sides, irregular, 8 corners

e octagon, 8 sides, irregular, 1 pair of parallel sides

Extended practice

1 Teacher: Several different ways of dividing the shapes are possible. The most likely are given below. Students' descriptions of the shapes will vary. Look for students who show a sophisticated understanding of shape and who can use a variety of criteria to describe the shapes in a way that makes them easily recognisable.

a 2 trapeziums **b** square, triangle

c 2 rectangles **d** triangle, trapezium

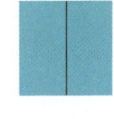 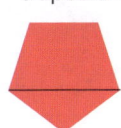

2 **a & b** Teacher to check. Teacher: Look for students who can combine the shapes into a new polygon.

3 Teacher: Answers will vary depending on the shape made. Look for students who can accurately name and describe the new shape they made using a range of criteria.

UNIT 6: Topic 2

Guided practice

1

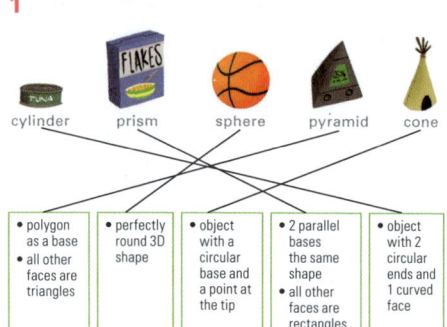

cylinder prism sphere pyramid cone

| • polygon as a base • all other faces are triangles | • perfectly round 3D shape | • object with a circular base and a point at the tip | • 2 parallel bases the same shape • all other faces are rectangles | • object with 2 circular ends and 1 curved face |

2

Independent practice

1 a A C D G

b D C G A

c Teacher to check. Teacher: Look for students who can make a reasonable attempt at drawing a 3D shape, and who recognise the faces of a square prism are all square and the same size.

d a cube

2 a

b

c

d

3 a–d Teacher to check. Teacher: Look for students who can identify mathematical similarities or differences, such as the shape of faces or the number of edges, rather than other cosmetic differences such as colour.

Extended practice

1

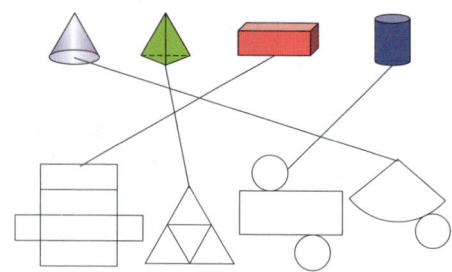

2 a Teacher to check. Teacher: Look for students who demonstrate an understanding of what a prism is, and who can identify the shapes that make up their object.

b Teacher to check. Teacher: Look for students who can use the features of their prism to accurately name it.

c Teacher to check. Teacher: Look for students who show a solid understanding of the features of 3D shapes and can write an accurate description that matches their sketch.

UNIT 7: Topic 1

Guided practice

1 a smaller   **b** smaller
c larger   **d** smaller
e larger   **f** larger

Independent practice

1 Teacher to check. Teacher: Look for students who show an understanding of right angles by finding and accurately representing items in the classroom that include them.

2 The following shapes should be circled: a, e, f

3 a 4   **b** 1   **c** 0

4 a 3 o'clock, 9 o'clock
b C, D   **c** B, F

5 Teacher to check. Teacher: Look for students who understand how to indicate an angle, and who can accurately classify the size of the angle in relation to a right angle.

Extended practice

1 Teacher to check. Teacher: Look for students who can apply their knowledge of angle sizes to successfully select and classify angles within the classroom.

2

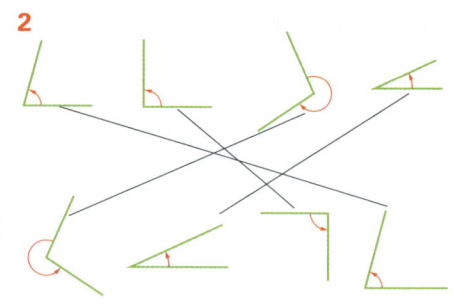

UNIT 8: Topic 1

Guided practice

1 a symmetrical   **b** symmetrical
c not symmetrical
d not symmetrical
e symmetrical   **f** symmetrical

Independent practice

1 In some cases, more than one answer is possible. The most likely responses are shown here.

a

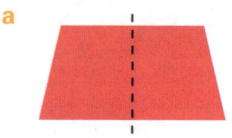

b

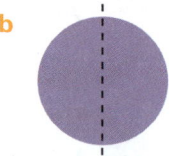

c

d

e

f

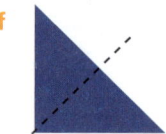

2 Teacher: Some of the shapes have more than two lines of symmetry. The most likely responses are shown, but accept any correct responses.

OXFORD UNIVERSITY PRESS

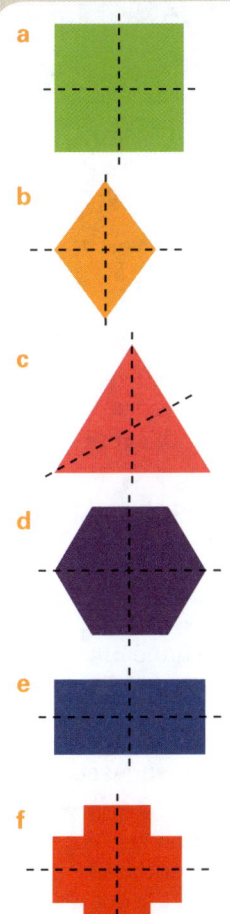

3 **a** Triangle or Shape c.

b Square, diamond, rectangle and cross or Shapes a, b, e and f.

4 **a & b** Teacher to check. Teacher: Look for students who can identify symmetrical items in the environment, and who demonstrate an understanding of symmetry in their representations of items and their lines of symmetry.

5 **a, c** and **d** should be circled

Extended practice

1 Teacher to check. Teacher: Look for students who can apply their knowledge of symmetry to make a simple picture that has either horizontal or vertical line symmetry.

2 Teacher to check. Teacher: Look for students who can demonstrate an understanding of line symmetry as two halves that are a reflection of each other.

UNIT 8: Topic 2

Guided practice

1 **a** slide **b** slide
c turn **d** turn

Independent practice

1 **a**

b

c

d

e

2 **a & b** Teacher to check. Teacher: Look for students who can apply their understanding of slides and turns to create their own pattern and accurately identify the rule.

3 **a** flip **b** slide **c** turn

4 Teacher to check. Teacher: Look for students who show awareness of translations in their environment and who can accurately represent and label their patterns.

Extended practice

1 Students may not identify all the translations present in each design.

a Pattern contains turns and flips.

b Pattern contains slides, turns and flips.

c Pattern contains slides and turns.

d Pattern contains slides, turns and flips.

2 Teacher to check. Teacher: Look for students who are able to demonstrate an understanding of translations and who can apply it to making their own designs.

UNIT 8: Topic 3

Guided practice

1 **a** a wombat

b a dingo

c a bird or rosella

d a crocodile

e a platypus

f koalas

Independent practice

1

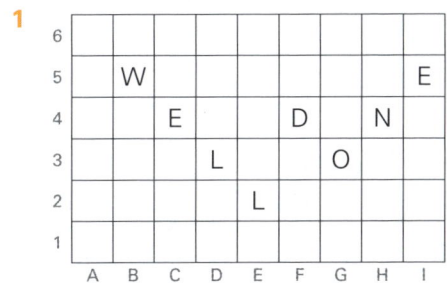

2 **a** B1 **b** A2
c D2 and E2
d A4, A5, B4 and B5
e F1 and F2 **f** D4 and D5

3 **a** Giraffe Road and Tiger Street
b Cat Road and Fish Road

4 Responses may vary – e.g. outside the shopping centre, on the corner of Dog Road and Goat Street, opposite the swimming pool.

5 Teacher to check. Teacher: Look for students who can use the language of direction to accurately navigate between the given points.

Extended practice

1 Teacher to check. Teacher: Look for students who can apply their knowledge of representing places on maps, incorporating features such as paths, buildings and trees, to make a map that is reasonably accurate.

2 Teacher to check. Teacher: Look for students who demonstrate an understanding of the language of direction by formulating accurate directions based on their map.

3 **a** B5, C1 or E4
b E3 **c** E1 **d** C3

UNIT 9: Topic 1

Guided practice

1

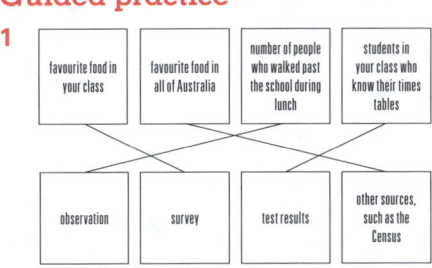

2 **a–c** Answers will vary.

Independent practice

1 **a** Answers will vary. Teacher: Accept any question that results in responses that can be categorised – e.g. "What is your favourite hobby?" or "Do you have any hobbies?"

b Teacher to check. Teacher: Look for students who successfully identify the categories for their data and who can accurately record their classmates' responses.

2 Question c should be circled.

3 Teacher to check. Teacher: Look for students who can accurately record 5 responses in the table.

4

Colour	Responses
Blue	IIII
Red	ЖII
Green	II
Pink	I

5 a Teacher to check. Teacher: Look for students who can frame an appropriate survey question to elicit a response that can be categorised – e.g. "What is your favourite animal?" rather than "What is your favourite animal like?"

b Teacher to check. Teacher: Look for students who can list the answers accurately and who have exactly 12 responses listed.

c Teacher to check. Teacher: Look for students who can identify appropriate categories for their data and who can accurately transfer the data from their list into the table.

Extended practice

1 a Teacher to check. Teacher: Look for students who identify that the categorical variable is the number of sides of the shapes.

b observation

2 a Teacher to check. Teacher: Look for students who recognise data that can be easily categorised through observation – e.g. the number of people in the class who wear glasses.

b Teacher to check. Teacher: Look for students who can categorise their data appropriately and record their data observations accurately in tabular or list form.

UNIT 9: Topic 2

Guided practice

1 a Favourite icy pole flavours in 3P
b Flavours
c Number of students
d 4   **e** 8
f Lemonade

Independent practice

1 a

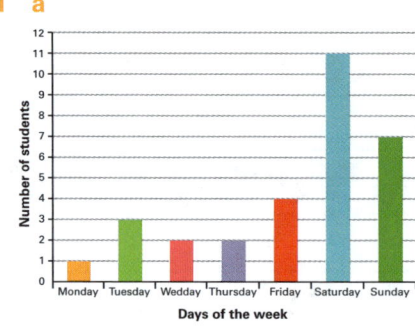

b Saturday   **c** Monday
d Days of the week
e Number of students
f 11

2 a & b Teacher to check. Teacher: Look for students who can collect and record data accurately in list form, and then translate that data to a pictograph.

c & d Teacher to check. Teacher: Look for students who can draw simple conclusions from their data.

3

	Country			
	Italy	**NZ**	**Australia**	**Vietnam**
No. of people	I	IIII	ЖII III	II

Extended practice

1 a–c Teacher to check. Teacher: Look for students who demonstrate an understanding of the data gathering process in the form of tables, pictographs and bar graphs by accurately depicting the same data on each.

d Teacher to check. Teacher: Students are likely to use a title such as "Position in family in 3N". Accept any titles that accurately reflect the data. The *y*-axis and pictograph label should indicate number of students, while the *x*-axis label should show position in family, or similar.

e Teacher to check. Teacher: Look for students who can use the language of statistics to justify their choice – e.g. the numbers on the *y*-axis of a bar graph make it easier to work out how many people are in each category, or the data in a pictograph gives you a quick visual of the results.

UNIT 9: Topic 3

Guided practice

1 a Interesting
b Fun   **c** Hard
d Boring, Challenging   **e** 26

Independent practice

1 a 7 **b** Cookies

c Teacher to check. Teacher: Look for students who can suggest plausible alternatives for the category – e.g. cake or carrot sticks.

2 Teacher to check. Teacher: Look for students who can make more sophisticated observations by comparing different parts of the data, such as the result in one category against the other, or aggregative data, such as recognising how many students were surveyed or the total of the two most favoured responses.

3 a Teacher to check. Teacher: The most likely responses are labels and numbers/scale, however accept any reasonable observation.

b Teacher to check. Teacher: Look for students who understand that a pictograph gives a quick visual snapshot of data, but that it is harder to use if numbers are required, as you have to count each item.

c Teacher to check. Teacher: Look for students who understand that bar graphs are helpful when you want to know exact numbers, especially when larger numbers are involved, as you can use the scale to quickly find the numbers for each category.

d Teacher to check. Teacher: Look for students who demonstrate that they can accurately interpret data and use it to draw conclusions.

e 8 **f** 19

Extended practice

1 a Teacher to check. Teacher: Look for students who can choose a topic that is appropriate for their age group, and who can formulate an appropriate question for their research.

b Teacher to check. Teacher: Look for students who can use appropriate methods such as lists or tables with tally marks to accurately track the responses to their surveys.

c Teacher to check. Teacher: Look for students who can construct a bar graph or pictograph that accurately reflects the data that they gathered.

d Teacher to check. Teacher: Look for students who can use their data to draw conclusions. More sophisticated responses may involve aggregating or comparing variables within their data.

Unit 9: Topic 4

Guided practice

1 a

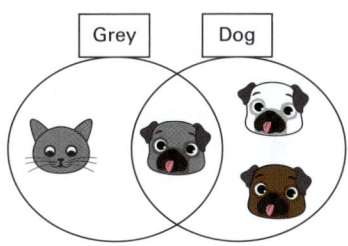

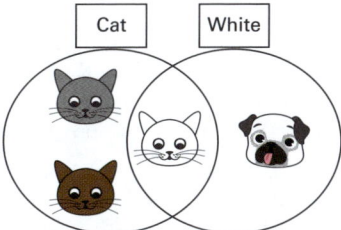

b

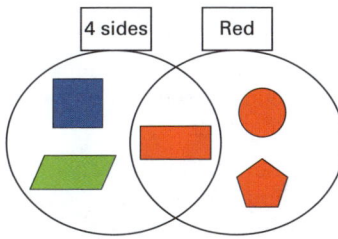

Independent practice

1 a

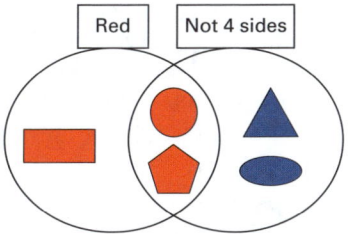

b triangle and oval

c rectangle

2

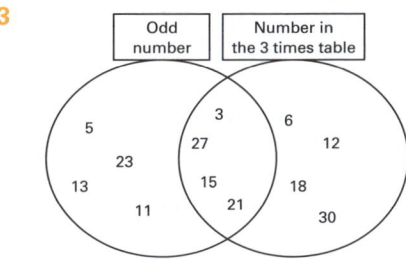

	Red	Not red
Curved	●	●
Not curved	⬠ ▬	▱ ▲ ▪

3

Odd number / Number in the 3 times table

5, 23, 13, 11 | 3, 27, 15, 21 | 6, 12, 18, 30

4 heads/heads, heads/tails, tails/heads, tails/tails

5 a

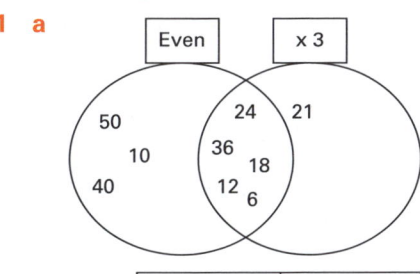

Possible outcomes
- red sock and red sock
- red sock and blue sock
- blue sock and red sock
- blue sock and blue sock

b Student circles: the same as a red pair.

c one quarter

d Answers may vary. Most likely response is because there are *two* outcomes that result in an odd pair but only *one* outcome for a blue pair.

Extended practice

1 a

Even / x 3

50, 10, 40 | 24, 36, 18, 12, 6 | 21

		x 5		x 3	
x 2	50	10	40	24 36 6 12	18
x 7	35			21	

b Answers may vary, e.g. 30.

c Answers may vary, e.g. 9.

2 a

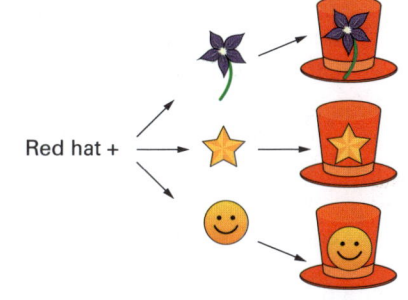

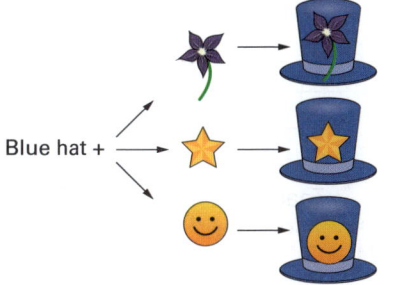

b There are 3 x 3 = 9 combinations.

c 36 ÷ 9 = 4 hats of each type are likely to be the same.

UNIT 10: Topic 1

Guided practice

1 a Teacher to check. Teacher: Look for students who make a reasonable estimate that is more than the result with only two flavour options, and who can justify their estimate using mathematical reasoning.

b Teacher: Accept any specific flavours or toppings students choose, as long as they fit into the categories below.

flavour 1 with topping 1, flavour 2 with topping 1, flavour 3 with topping 1, flavour 1 with topping 2, flavour 2 with topping 2, flavour 3 with topping 2

c 6

Independent practice

1 **a** red and blue, red and green, red and yellow, blue and green, blue and yellow, green and yellow

b Teacher to check. Teacher: Look for students who recognise that the addition of another colour will result in more possible outcomes.

c red and blue, red and green, red and yellow, red and purple, blue and green, blue and yellow, blue and purple, green and yellow, green and purple, yellow and purple

d 10

e less likely

f impossible

2 **a** 4

b **i–iv** Teacher to check. Teacher: Look for students who recognise that red is the most likely and blue and green the least likely colours, and who choose appropriate chance words to reflect this.

c red

d blue and green

3 **a–d** Teacher to check. Teacher: Look for students who have more green segments than any other colour, fewer blue segments than other colours, no yellow segments, and more red segments than blue. It is acceptable for students to use other colours as long as the criteria are met.

4 **a** 2 **b** 4 **c** 8

5 Teacher to check. Teacher: Look for students who show an understanding that there is an equally likely chance of tossing heads or tails, and therefore coin tossing can be a fair way to make simple decisions when people cannot agree.

Extended practice

1 **a** Teacher to check.

b red, blue, pink
red, pink, blue
blue, pink, red

blue, red, pink
pink, red, blue
pink, blue, red

2 Teacher to check. Teacher: Look for students who show an understanding of a range of chance phrases, and who can accurately apply these to the situation presented.

UNIT 10: Topic 2

Guided practice

1 **a** Teacher to check. Teacher: Look for students who make reasonable predictions that encompass a spread of numbers, and who can appropriately justify their answers.

b Teacher to check. Teacher: Look for students who can accurately record the outcomes – e.g. there should only be 10 results listed.

c Teacher to check.

d Teacher to check. Teacher: Look for students who show an awareness of the role that chance plays in the experiment and who use reasoning to justify why their results may not have been as expected.

Independent practice

1 **a** Teacher to check. Teacher: Look for accurate recording of exactly 30 outcomes.

b Teacher to check. Teacher: Look for students who demonstrate an understanding of the randomness of chance, and who can use the language of probability to support their assertions.

c Teacher to check. Teacher: Look for accurate recording of exactly 30 outcomes.

d Teacher to check. Teacher: Look for students who focus on the chance element when comparing data and who show that they understand that the dice could land on any number each time.

e Teacher to check. Teacher: Look for students who demonstrate an understanding of the difficulty of accurate predictions when chance is involved.

f Teacher to check. Teacher: Look for students who are understand that there is a smaller likelihood of each number being rolled when using a 10-sided dice than a 6-sided dice.

2 **a** heads, tails

b tails/tails, tails/heads, heads/tails, heads/heads

c equally likely

d Teacher to check. Teacher: Look for accurate recording of exactly 20 outcomes.

e Teacher to check. Teacher: Look for students who can accurately interpret their results to identify the most frequent outcome.

f Teacher to check. Teacher: Look for students who can accurately interpret their results to identify the least frequent outcome.

g The ideal response is "no". Teacher: Look for students who demonstrate an understanding of the role of chance in the results and therefore expect differences between their own and others' results.

h Teacher to check. Teacher: Look for students who understand that chance means results are unlikely to be the same two times running.

3 Answers will vary. Students are most likely to circle "winning a raffle" and "catching a cold"; however, the other answers are acceptable if students can adequately justify their choices, e.g. your chance of getting a perfect score on a spelling test might be influenced by the words you are being tested on.

Extended practice

1 **a** Teacher to check. Teacher: Look for students who recognise that chance will determine which colour is drawn out and it is therefore difficult to predict the colour with any accuracy.

b Teacher to check. Teacher: Look for accurate recording of exactly 25 outcomes.

c Teacher to check. Teacher: Look for students who can accurately translate the results of their chance experiments into a graph.

d Answers will vary depending on student data. Teacher: Look for students who are able to accurately interpret their results using the language of chance.

OXFORD UNIVERSITY PRESS